The Beast Within

ANIMAL WORLDS

Jessica Serra, Series Editor

What if,
instead of looking at animals
through our own eyes,
we looked through theirs?

Recent scientific discoveries offer us a new perspective on the animal kingdom, shattering the myth that once equated animal behavior with that of machines. We now know that humans are not the only beings with intelligence, emotions, and language skills.

Even though animals share our environment, they perceive and understand it in their own way. Equipped with specific sensory equipment, they selectively pick up certain meaningful signals and evolve in a world of their own. This means that our human world is only one among millions of others.

Shifting our perspective to reflect this reality forces us to rethink our own place in the world, not as superior to other living beings but amid them. This perspective also allows us to discover the infinite richness of animal lives and the dazzling complexity of "beasts."

Enlightened by science, this series endeavors to open doors to these other worlds by providing a new understanding of living things and, therefore, a new understanding of ourselves.

The Beast Within

HUMANS AS ANIMALS

Jessica Serra

Translated by Alison Duncan

Johns Hopkins University Press

BALTIMORE

© 2024 Johns Hopkins University Press

All rights reserved. Published 2024

Printed in the United States of America on acid-free paper

9 8 7 6 5 4 3 2 1

This work was originally published in French as *La bête en nous* © humenSciences/ Humensis, 2021.

Johns Hopkins University Press

2715 North Charles Street

Baltimore, Maryland 21218

www.press.jhu.edu

Cataloging-in-Publication Data is available from the Library of Congress.

A catalog record for this book is available from the British Library.

ISBN: 978-1-4214-4750-6 (paperback)

ISBN: 978-1-4214-4808-4 (e-book)

Special discounts are available for bulk purchases of this book.
For more information, please contact Special Sales at specialsales@jh.edu.

For Jérôme

CONTENTS

The Beast Within

1

Denying
Our Animality

Man's derived supremacy over the earth; man's power

of articulate speech; man's gift of reason; man's free will

and responsibility; man's fall and man's redemption—

all are equally and utterly irreconcilable with the

degrading notion of the brute origin of him who

was created in the image of God.

SAMUEL WILBERFORCE,
"IS MR. DARWIN A CHRISTIAN?," *QUARTERLY REVIEW*, 1860

"Is it through your grandfather or your grandmother that you claim your descent from a monkey?" Thus began, in June 1860, the legendary heated debate—also known as the "Great Debate"—between the English bishop Samuel Wilberforce and his compatriot, the paleontologist Thomas Henry Huxley. Huxley replied that he was not ashamed to have an ape as an ancestor, but that he would be if his ancestor had been a clergyman dealing with scientific matters about which he knew nothing. Voices immediately erupted within the University Museum, protesting this offense against religion. In the uproar, Vice Admiral Robert FitzRoy was seen brandishing a Bible and imploring the audience to "believe God rather than man." How could we even consider that primates could be our ancestors?

Preposterous! And repulsive. As a matter of fact, Huxley was the faithful friend of Charles Darwin, the English naturalist who had studied theology for many years and whom FitzRoy also knew very well. At the young age of twenty-two, Darwin had been invited on FitzRoy's ship, the *Beagle*, to join the great scientific expedition. FitzRoy, a meteorologist and leading specialist in topographic surveys, had been looking for an educated traveling companion to keep him company and had asked several of his acquaintances, all of whom had declined the offer. It was finally the young Charles, who came highly recommended by his peers, who fit the bill. He had just finished his studies in theology and enthusiastically prepared his gear for his scientific adventures, with the intention of bringing back many specimens.

This incredible expedition began one day in December 1831 and lasted almost five years. Three of those years were spent on land, during which time Darwin tirelessly explored many territories, including Argentina, Patagonia, the Atacama Desert, and the Andes Mountains. On the Galápagos Islands, the young man noticed that certain species of finches that closely resembled one another were unable to reproduce together. He quickly understood that each species had characteristics adapted to the environment in which it had evolved. That was when his brilliant mind allowed him to develop a foundational hypothesis that undermined both creationism as told in the Bible and the theory of the fixity of species. According to his hypothesis, animals must have had a common ancestor who had changed over time, giving rise to the multitude of species we know today. Revolutionary! This adaptive transformation of species took place through natural selection. Darwin described this process, writing, "The favourable variations would tend to be preserved, and unfavourable ones to be destroyed. The result of this would be the formation of new species."[1] But at the time, to avoid criticism, the young Darwin chose to keep his idea to himself. It was not until over twenty years later, in 1859, that he decided to put his theory down on paper in his book *On the Origin of Species*,[2] a work whose first print run sold out in a single day. So, in 1860, when FitzRoy found himself face to face with Huxley, who was supporting the thesis of the man he had taken under his wing twenty-five years earlier, his blood boiled. As a fervent believer, FitzRoy resolutely defended "the argument from design," which claimed that the perfection of organic structures proved the hand of God. The vice admiral could never have guessed that the young man with whom he had had so many conversations, the very one he had welcomed on his famous ship to keep him company, would one day sound the death knell for creationism.

For the first time in millennia, Darwin reinstated humankind's place within the animal kingdom. His theory of evolution, which instantly erased the boundary separating humans from animals and thrust *Homo sapiens* into a vortex of millions of species, created immense unease. Contrary to popular belief, this scientific precept not only claimed that we descended from apes but that we are in fact apes, and "naked" apes at that, as the English zoologist Desmond Morris pointed out in one of his books.[3] We can all call to mind the famous drawing published in 1965 in which a primate gradually becomes more erect over time until finally becoming modern man. But we must be careful how we interpret that: it does not mean that modern man descended in a straight line from an ancestor shared with other primates but rather that man descended from an entire lineage of apes and is the result of numerous instances of crossbreeding. The transition from walking on all fours to standing upright makes it appear that there was a destination, as if all evolutionary change followed a linear upward progression culminating in a final version of the modern human. But that is not the case. Evolution is fluid and unpredictable.

Instead of picturing a two-dimensional evolutionary tree with continuous lines, as is still common to find in biology books, we should envision a three-dimensional universe in which the tree of life never stops growing and has millions of branches going in thousands of directions, intertwining in places and stopping abruptly in others because of an extinction event. Each little bud represents a species that will grow into a new branch of its own. Humans do not sit at the end of the prettiest branch at the tip-top of the tree. Our spot is a small bud positioned on a twig similar to all the other twigs and, just like all the other small buds, our species is constantly changing and continues to evolve. Our genes are constantly interacting with our environment, which means that what we consume, our exposure to pathogens, climate conditions, pollution, and many other factors impact our genes. Scott D. Solomon, a biologist at Rice University in Texas, explains that human sequencing data provides researchers with proof that natural selection continues to affect our genes. And this does not happen slowly over the course of a millennium or a century, but at the much faster pace of within a generation. For example, the DNA of a newborn contains on average about sixty mutations when compared with the parents' genomes. Our children are not a mere combination of our genes; they are a new version of a human being.

As an example, depending on the genes we inherited from our ancestors, some of us can easily digest milk as adults while others have a lot more difficulty. In prehistoric times, once weaned, none of our ancestors was able to metabolize milk. Certain genes stopped producing the enzyme necessary for digesting it, because after being weaned around the age of three to four years, the young no longer had access to mother's milk, which was only produced during lactation. Conditions changed nine thousand years ago when humans began to domesticate certain animals (cows and sheep in particular). Consequently, humans no longer relied on mother's milk, and even adults could drink milk from their animals, a beverage whose nutrient content greatly increased their chances of survival. The genome of some individuals underwent changes that allowed them to continue to produce this enzyme after weaning. This meant that these individuals had a selective advantage, and they transmitted this mutation to their offspring. These humans were thus able to drink milk their entire lives. Today, this mutation is present in a quarter of the world's population, with a strong disparity between northern Europeans and North Americans, who for the most part consume milk products without difficulty, and the great majority of Africans, South Americans, and South Asians, who tend to be lactose intolerant.

The environment molds our genome over generations. An example from research in Argentina illustrates how this happens. A team of Swedish researchers discovered an Argentinian population better able than other humans to withstand high concentrations of arsenic in drinking water.[4] In the city of San Antonio de los Cobres, the concentration of this toxic compound is twenty times higher than the limit set by the World Health Organization (WHO). If Europeans were to drink this water, they would show severe signs of poisoning. By comparing this Argentinian population with other South American populations, Carina M. Schlebusch, a professor of evolutionary biology at Uppsala University in Sweden, and her colleagues were able to prove the existence of a variation in a gene involved in metabolizing arsenic, which is how our bodies break down and excrete this toxic substance. She thus discovered that the population of San Antonio had developed a resistance to the metalloid. In fact, the human genome is constantly changing through a series of mutations occurring in each generation, meaning that the modern human is not the final stage in an evolutionary process. Humans continue to be subjected to the laws of evolution.

Considering that humans are animals by nature and are constantly changing, defining what is unique to human beings is an arduous task. Since the humans of today are not the same as those of yesterday or those of tomorrow, questions arise about how to define human nature. Since human characteristics change over time, does that make human nature a relative concept?

Is Humanity a Relative Concept?

If we could board Dr. Emmett Brown's DeLorean (the time machine from the film *Back to the Future*) and be transported to the dawn of humanity, say around 3 million years BCE, we would see that nothing differentiated the first humans from other animals. Births were difficult. Humans entered the world on the bare ground or on a pile of leaves. Most died before even learning to walk. Sometimes they were too weak to withstand disease or extreme cold, or they were devoured by other animals. Those who survived to adulthood spent much of their time searching for food, making friends, fighting over territory, and were in constant fear of predators. Even so, they had a range of emotions probably very similar to what we have today. These primates had the capacity to love, hate, be afraid, experience pleasure, and be sad. But sooner or later, like all animals, they eventually died, leaving no trace of their time on this earth except some dust or a few fragments of bone. As Pascal Picq, a paleoanthropologist and former professor at the Collège de France, reminds us, the first humans, somewhere in Africa around three million years ago, were not pondering the human condition while climbing down from their tree each morning or before leaving in search of carrion in the savanna woodland.[5]

Even though today there is no consensus on the date "true humans" (in the sense of the first primates capable of abstract thought) emerged, scientists do agree that a major change event took place two million years ago around the time *Homo erectus* appeared. It was then that the first modes of symbolic expression developed, particularly cosmetic ones (modifying one's appearance with makeup, tattoos, or scarification), as well as those related to cosmogony (first beliefs and the birth of spirituality).[6] Along with the development of linguistic, artistic, and symbolic skills came the acquisition of abstract thought. The appearance of funeral rites revealed the emerging questioning about the meaning that should be given to death and the desire to no longer be easy prey—a *piece of meat*—for predators. For the first time in their history, humans wanted to escape their mortality. Many anthropologists see this as the birth of humanity, because our ancestors began to acquire an awareness of the cosmos. Judeo-Christian religions are based on the precision that humans are higher life forms than animals, raising humanity up all alone above a bestial mass devoid of intelligence and emotions. The latest anthropological discoveries, however, show that forty thousand years ago our *H. sapiens* ancestors were not the only species of humans on Earth. As Mark G. Thomas, an evolutionary geneticist at University College London, underscored, "What it begins to suggest is that we're looking at a *Lord of the Rings*-type world—that there were many hominid populations."[7] While no Ringwraiths or Elves are mentioned in our evolutionary history, our predecessors shared the earth with creatures whose bodies were just as varied as those described by J. R. R. Tolkien.

So still aboard our DeLorean, if we travel to a time forty thousand years ago, we would come across Neanderthals, Denisovans, Flores Men, and probably many other species of humans. Over the past few years, the image we once had of Neanderthals (as big stocky brutes devoid of intelligence, with prominent brow ridges and large jaws) has also evolved. Researchers have shown that these prehistoric humans had a brain 10% larger than *H. sapiens'* and, like modern humans, possessed the gene crucial for modern speech.[8] Additionally, Neanderthals had a bone located in the anterior part of the neck whose biomechanical characteristics and structure resembled *H. sapiens'* hyoid bone, implying that they had the capacity for articulate speech and complex language.[9] These robust beings had a large rib cage and short legs but did not move with their knees constantly bent like apes; they stood upright with a range of motion similar to that of *H. sapiens*. From excavation after excavation, paleontologists have discovered that Neanderthals fully mastered fire, wore clothes and jewelry, built all kinds of huts, took care of the weak, and buried their dead. Better yet, Neanderthals could picture objects in their minds before making them. The archaeologist Marie-Hélène Moncel, director of research at the French National Centre for Scientific Research (CNRS), found proof that Neanderthals were skilled at weaving cords. She discovered evidence at the Abri du Maras in the Ardèche department of France in the form of twisted plant fibers on stone tools.[10] In addition to the cognitive ability to learn through repeated trial and error, this type of cord production requires a mathematical understanding of pairs, sets, and numbers, and an agility of the fingers far superior to that of *H. sapiens*. Furthermore, the tools that Neanderthals made took hours to produce and required that they thought through the steps chronologically beforehand. And that's not all. New dating methods developed by Dirk L. Hoffmann's team at the Max Planck Institute for Evolutionary Anthropology in Leipzig, Germany, revealed that some of the wall paintings found in caves were not made by *H. sapiens* but by Neanderthals.[11] They were much more sophisticated beings than we had once thought. According to the zoologist and paleoanthropologist Clive Finlayson, who conducted research on his home island of Gibraltar, the pattern found in Gorham's Cave (crisscrossing horizontal and vertical lines) is also the handiwork of Neanderthals, who occupied this geographical area forty thousand years ago. Contrary to the ubiquitous cliché that gave us the expression "caveman," these humans did not live in caves but used the surfaces of cave walls to express themselves artistically. The recent reconstruction of a Neanderthal girl's face from Gibraltar completed by the Anthropological Institute at the University of Zurich showed that her light-skinned face, soft features, green eyes, and light brown hair were a far cry from the image of the "beast" half ape, half human that we had once envisioned.

In examining the brain of Neanderthals, various studies have tried to estimate their intelligence by comparing it to *H. sapiens'* brain. In 2010, researchers at the Max Planck Institute put forward the hypothesis that *H. sapiens* are intellectually superior because of the fact that their brain size increases during the first years after birth, which did not happen in Neanderthal children.[12] But new findings finally showed the opposite to be true: brain-development processes are broadly similar in Neanderthal and modern human children.[13] The debate was not settled, however, because another team of scientists, this time from Japan, revealed neuroanatomical differences in the cerebellum through a three-dimensional reconstruction of these two species' brain anatomy. The scientists claimed that these differences explained a lack of behavioral flexibility in Neanderthals and, therefore, a less developed intelligence.[14] Nevertheless, Jean-Jacques Hublin, director of the Department of Human Evolution at the Max Planck Institute, has since qualified the conclusions of this Japanese team by reminding us of the fundamental fact that there is not just one type of intelligence. "A person can have a very underdeveloped social intelligence, and yet perfectly master fields as complex as quantum physics," he explained.[15]

How did Neanderthals perceive their relationship to the cosmos? To answer this question, the geneticist George Church at Harvard University announced that he wanted to bring this ancient human back from the dead by reconstructing its DNA using samples from bone remains.[16] Once reconstituted, the DNA would be put into stem cells and then injected into the cells of a human embryo, which in turn would have to be implanted in the womb of a surrogate mother. There were still a few obstacles to overcome though, starting with finding a woman willing to volunteer to carry a Neanderthal baby. Faced with criticisms that evoked the Franken-stein syndrome, the scientist retorted that this future being might think differently than we do and would be an exceptional opportu-nity to increase human diversity. He argued that this could prove valuable for the survival of our species because "if you become a monoculture, you are at great risk of perishing."

Until Neanderthals rise from the ashes, we can only speculate on what they were like based on archaeological remains. For this prehistoric human, life expectancy was short, and few children reached adulthood. Some of them were eaten alive, for instance by hyenas or cave bears, while survivors looked on in terror, helplessly watching a friend, a sister, or a baby be dismembered. Predators even ingested Neanderthals' heads, as illustrated by the discovery of partially digested teeth and skull fragments at the Mousterian site of Marillac-le-Franc located in the Charente department in France.[17] Death lurked incessantly; living in groups was the only way to sur-vive in this prehistoric world. These hunter-gatherers, who fed partly on plants, found consuming animals to be a welcome source of protein to feed their energy-consuming brains. They then took on a predatory role of their own, and not just any role. Neander-thals became specialized in tracking and killing giant herbivores, such as woolly rhinoceroses and mammoths.[18] Armed with simple spears, Neanderthals moved in small groups and required several attempts to finish off their prey, inflicting wound after wound until the animal succumbed. Death was part of their daily life.

But Neanderthals were far from being indifferent to losing one of their own. In fact, they buried the bodies of their deceased clan members so as to protect them from the ferocious appetites of scav-engers. They even invented burial rituals. Neanderthals built tombs using stone slabs and laid the deceased on beds of flowers or accom-panied by stone tools. It was in Shanidar Cave, located in the Erbil

Governorate of Iraqi Kurdistan, that this previously unknown "flower burial" rite was discovered. Fossilized pollen found at the very place where one of the ten dead bodies had been buried suggested that Neanderthals had a ritual of placing "floral offerings." The palynological study (analysis of the pollen) revealed that the flowers had been whole when they were delicately deposited in the tomb. This detail is of paramount importance because it proves that the placement of plants or objects near the corpse was intentional. Neanderthals were capable of conceiving of an invisible world, of the "beyond."

We know less about the habits and customs of Denisovans, our other cousins. Molars found near a superbly crafted chlorite bracelet along with a needle with an eye for sewing clothes suggest that Denisovans' mental faculties were similar to Neanderthals'. Flores Men in Indonesia looked more like Tolkien's hobbits. This species likely evolved in isolation over the course of thousands of years, which would explain its extremely small body size better adapted to island conditions. There is little certainty about the cognitive ability of the Flores Man. Its brain was no larger than a grapefruit, much smaller than that of its ancestor *H. erectus* for its body size. But archaeological discoveries have led us to believe that this species had also mastered fire and tool making, skills that would have been retained by cultural transmission despite the progressive shrinking of its brain size.

Did *H. sapiens* ever meet these ancient humans? In 1957, a team discovered a mandible in Riparo Mezzena, a rock-shelter in Italy, with unusual characteristics. The jawbone appeared to belong to a Neanderthal, but it also had a prominent chin, a defining feature of modern humans. It was not until 2013 that researchers were able to determine that this jaw belonged to a hybrid, born to a Neanderthal mother who had mated with a *H. sapiens* father.[19] Our ancestors therefore interbred with other species at various times throughout their history. But the genetic differences between the two species were such that only unions between a male *H. sapiens* and a female Neanderthal could produce viable and fertile children. The inverse would end in miscarriage or produce offspring who were unable to reproduce as adults. In fact, the researchers found no trace of Neanderthal DNA on the Y sex chromosome of *H. sapiens* (present only in male individuals). *H. sapiens* also interbred with Denisovans and were able to give birth to fertile children. It cannot be ruled out that *H. sapiens* also tried to mate with Flores Men, but it is unlikely that there were descendants because their extremely small size would not have facilitated childbirth or the birth of strong offspring.

We cannot know if this interspecies mating was due to rape or genuine relationships, but it was more or less successful and resulted in around 4% of Neanderthal genes and 1% of Denisovan genes remaining present in our genome. Accordingly, we always carry within us a bit of these *other* humans. Unfortunately, this is the only thing we have left since Neanderthals, Denisovans, and Flores Men died out soon after their encounter with *H. sapiens*. Even though Neanderthals had been the undisputed dominant species in Europe for hundreds of thousands of years, *H. sapiens*' arrival in the same territory forty-five thousand years ago marked the beginning of the end for Neanderthals. Some five thousand years later, Neanderthals found themselves isolated to France and the western part of the Iberian Peninsula. Another ten thousand years passed before the Neanderthals were definitively erased from the surface of the globe. There is little doubt that *H. sapiens* were responsible for their extinction, and probably the extinction of all other human species too. *H. sapiens* colonized lands that were not their own, and Neanderthals saw prey and edible plants become scarce with the arrival of this invader. The prehistorian Marylène Patou-Mathis, director of research at the CNRS, put forward the hypothesis that Neanderthals were faced with an insurmountable "stress," stating, "The arrival of different humans but who looked like them undoubtedly provoked more fear than curiosity. Rather than confront the intruders, they preferred to avoid encounters. But their population density was insufficient. Infant mortality increased and, in just a few generations, the population decline was able to get the better of the last tribes."[20]

Many see *H. sapiens'* world domination as proof of their exceptional intelligence. As we have seen, however, many other human species endowed with strong cognitive abilities traveled the world before them. Neanderthals, for example, had their own intelligence. It was certainly different from that of our ancestor but not necessarily inferior. Both were capable of solving complex problems. What happened, then, to make *H. sapiens* win the evolutionary battle?

As an ethologist, the most beautiful expressions of intelligence that I have been able to observe in animals are the outcomes of collective intelligence. The importance of this phenomenon is made apparent to us while gazing upon the architecture of termite mounds, for example. These bioclimatic towers that measure several meters high make use of solar heating to ventilate the air inside and achieve a consistent temperature in the nest. These emergent structures arise from decentralized local interactions. We call this "self-organization" or "spontaneous order." My conviction is that, in order to reach the level of technological and scientific progress that we know today, *H. sapiens* have their own version of self-organizing behavior. This explains why our species is capable of sending a robot to Mars, while at the individual level, the vast majority of us have no understanding of the laws of mechatronics. We need what I call the "melting pot of brains," which is to say the pooling of a certain amount of knowledge and know-how that is transmitted from generation to generation and thus increases our collective intelligence.

I do not believe in the theory that *H. sapiens* differentiated themselves from other humans when certain individuals suddenly became capable of language and abstraction. Our ancestors were faced with major challenges that made them adopt cooperative techniques that were more effective. When they left Africa to conquer Europe, their encounter with Neanderthals was a pivotal moment. How could they compete with an endemic species that was physically more powerful than them? They had to be better hunters than Neanderthals, and, at several times in history, they had to know how to contend with the Neanderthals' formidable strength. In groups of equal numbers, *H. sapiens* certainly could not have competed with their cousin, either in hunting prey or in direct confrontations. The only solution was to outnumber them. This required transmitting a greater amount of information. Unlike social insects, whose genome plays a preprogramming role in the distribution of tasks (which is what enables such complex social structures to emerge naturally), *H. sapiens* had to *learn* to cooperate more effectively. This propelled the species into a higher level of social cognition. Developing empathy certainly helped, but above all they had to make sure information was passed down from one generation to the next in order for this connected social system to remain effective. The development of language, which at this time was still in basic form, was one means of achieving this. It allowed them to explain, coordinate, share action plans, and anticipate different scenarios. *H. sapiens* developed their language extremely quickly, and this made it feasible to share each discovery with as many people as possible. By pooling their knowledge, the newly created clans improved their hunting and combat techniques. They made use of their environment in a very different way from how other animals have throughout history. The more the groups became connected, the more individuals saw their own personal capacities grow. This was not due to some *genetic miracle*, but rather to inheriting a learned body of knowledge from their predecessors. In turn, this type of learning itself generated more and more complex mental phenomena, such as memorization. The growing complexity of language and abstract thought made it possible to create a collaborative network that was infinitely larger and better adapted than any other and that allowed *H. sapiens* to gain new resources.

From my point of view as an ethologist, it is the behavioral changes brought about by life in society that caused *H. sapiens'* brain to evolve in a unique way. By creating a safer and more stimulating environment and by acquiring through culture an ever-greater amount of knowledge with each generation, children whose brain plasticity was more developed and better adapted to absorbing a massive influx of information transmitted by their parents benefited from a selective advantage. They then gave birth to offspring whose brain architecture made it possible to absorb an even larger amount of knowledge. Recent anthropological discoveries also show that the brain shape of *H. sapiens* gradually changed from an elongated to a globular one, only evolving into its *modern* shape thirty-five thousand years ago, the date corresponding to when the Neanderthals disappeared.

This progression of changes is not the only element that favored *H. sapiens'* victory. The larger their groups grew, the more the wolves began to prowl near their camps, attracted by the smell of meat and the waste discarded nearby. This formidable predator fascinated our hunter-gatherer ancestor, who probably spent hours analyzing its hunting techniques, which were based on social cooperation. Endowed with an extremely developed sense of smell, canines can detect an herbivore more than 2 kilometers (nearly 1.25 miles) away. Their brain is able to map places with great precision. When hunger strikes, some members of the pack initiate the attack on a herd, and when their prey turns about-face, they find themselves face-to-face with the other wolves. Whether it was

H. sapiens' curiosity or their attraction to newborn mammals that prompted them to bring orphaned wolf pups back to their camps is unclear, but what matters is that this did happen. What might have been nothing more than a fleeting interest would ultimately change the course of humanity. The young wolf pups showed a strong attachment to the humans who raised them. An ethologist and director of research at the CNRS from Montpellier, France, named Pierre Jouventin recounts adopting a wolf pup himself "before she even opened her eyes" in his book *Kamala, une louve dans ma famille*[21] (Kamala: A wolf in my family). The pup was born in a zoo in 1976 and was scheduled to be euthanized. Instead, she grew up with the researcher, his wife, and his son in their apartment. In his book, the scientist describes an altruistic animal that would defend the humans she lived with as soon as she perceived any danger. Jouventin observed an innate behavior of familial mutual aid, which would explain both the social cohesion observed in wolf packs and also the reason that motivated *H. sapiens* to keep this canine by their side. Once tamed and grown, this powerful animal defended the humans with whom it lived, even if that meant putting its own life in danger. Moreover, wolves helped humans catch prey with unrivaled efficiency. Armed with their keen sense of smell and fast speed, the wolves' job was to wear out bison, mammoths, and rhinoceroses until they slowed down enough for the humans to then finish them off with spears or arrows. Humans immediately understood the benefits of their incredible alliance with wolves, which propelled this dynamic duo to the top of the food chain. Cave bears and hyenas had to watch out! From then on, *H. sapiens* and their wolves dictated the rules of the great chessboard of life.

As humans had already acquired the ability to pass along discoveries to as many others as possible, the domestication of wolves spread, and humans began selecting the canine individuals most receptive to orders. These animals spread at breakneck speed, accompanying *H. sapiens* on all their migrations. Through selection by *H. sapiens* over generations for its docility and aptitude for cooperation, and due to a phenomenon called "neoteny,"* the wolf showed more and more juvenile behaviors, which made it more adapted to its new life with humans. Playful behaviors characteristic of pups, such as barking or frantic tail wagging, remained in adulthood. Fed by humans and protected from the risks of predation, the wolf lost some of its adaptive specializations from its former life in the wild. Its snout, paws, and carnassial teeth shrank, along with its skeleton. The ferocious beast metamorphosed into a new species with the behavior of an eternal pup: the dog!

We can see that *H. sapiens'* success was not only due to their capacity for symbolic thought or the miraculous appearance of language but was also thanks to their aptitude for cooperation, which enabled them to make their discoveries in innumerable fields available to all. Their alliance with the wolf also played a considerable role in their expansion. Several species of humans with remarkable cognitive abilities, their own cultures, and their own beliefs lived alongside the wolf for thousands of years. They were not a failed version of modern humans. If they had borrowed the same strategies as *H. sapiens*, perhaps we would have Neanderthal or Denisovan neighbors today. We can also say that humanity, from an ontological point of view, is plural. It cannot be defined as characteristic of a species that would result from an ordered world. Each in their own way, several species of humans contemplated their "humanity."

Our Buried Animality

Today, when modern humans consider the concept of humanity, they give it substance by differentiating it from *nonhuman* behaviors.

---

* In developmental biology, "neoteny" is the conservation of juvenile characteristics in the adults of a species.

In this way, humanity only exists in opposition to beasts who act on instinct, and animality is the boundary humans transcended. We do not know much at all about the way Neanderthals and Denisovans imagined their own place in the universe, but as far as *H. sapiens* are concerned, many clues seem to reveal that in prehistoric times our ancestors had no need to regard themselves as being superior to other animals when they began thinking about their humanity. On the contrary, their philosophical explorations gave animals a central position. When *H. sapiens* engaged in abstract representations on rocks and cave walls, they could have drawn the sun, the moon, trees, or flowers, but they chose to only recreate animals, human hands drawn in stencil, or lines and dots. Without exception, all the paintings attributed to them present this distinctive feature. How can this overrepresentation of animals be explained? The majority of paleoanthropologists point to shamanism. According to this theory, since the dawn of *H. sapiens*, cave paintings were used in rituals to communicate with the *otherworld*. This otherworld would become entangled with the real world through dreams and visions, or when in a trance state caused by fasting or ingesting hallucinogenic plants. Some individuals (known as "shamans," but also others at pivotal moments in their lives) could come into direct contact with spirits by engaging in altered states of consciousness. Outlines of hands on the walls testify to this desire to create interstices between the real world and the invisible one, or as the French prehistorians Jean and Geneviève Guichard underscore, to link "the sacred and the profane, the human and the inhuman, the supernatural and the natural, the visible and the invisible."[22]

For *H. sapiens*, the otherworld was laden with powers. It influenced all life events, from the arrival of rain to the abundance of prey, hence the importance of communicating with the beyond. The caves were not chosen by chance: they symbolized the tunnels that led to this invisible dimension. Drawings of animals—whose aesthetic sophistication would make Picasso say millennia later, "We have invented nothing!"—were located where sound echoes strongly[23] to make human voices reverberate in the dark depths. The artists used the rock's natural relief to showcase their compositions, giving the impression that the works were rising from the stone. Thus, with great anatomical precision, these murals elicited a mystical feeling. This shamanic hypothesis, put forward by Jean Clottes, a French expert in cave art, and his colleague David Lewis-Williams, a South African archaeologist (coauthors of *The Shamans of Prehistory: Trance and Magic in the Painted Caves*),[24] is defended by various specialists. It therefore seems that in those ancient times, *H. sapiens* saw their relationship with animals as being horizontal and nonhierarchical. This theory can be verified by observing groups of contemporary hunter-gatherers who all share an animist belief system (the belief that spirits are incarnate in animals), even if each culture has its own rituals.

This primordial religion, which is the longest-observed religion of our ancestors, gave animals a central place. Cave walls, like a permeable veil, were the bridge between the visible and supernatural worlds that spirits could cross through the animal figures. The only depictions of human figures are hybrid beings, called "therianthropic deities," with a human body and an animal head. In the brains of hunter-gatherer *H. sapiens*, animals were the embodiment of spirituality, united with spirits by a common life force. The most feared predators were a symbol of power. Even hunting was an agreement made with animals, a kind of exchange of spirits. Humble before the forces of nature, our foraging ancestors did not feel the need to create a deity that would raise them to a status at the top of the world. They thought of their humanity as one of the multiple manifestations of the acting forces, just as animals were another.

Around 9 thousand years BCE, the wolf, which had become a dog through its life with humans, enabled an extraordinary population increase for *H. sapiens*, who no longer had to fear predators. Gone were the days when humans risked being torn apart by a pack of hyenas. Propelled to the top of the food chain, *H. sapiens* wiped out a large number of animal species from the surface of the globe. No mammoths, giant sloths, or woolly rhinoceroses survived the assaults of the dog and biped-human hunting team. When these species disappeared, a large part of their predators did too, as they were suddenly deprived of their prey and struck down by humans. The fearsome saber-toothed tigers, cave bears, and cave lions have since become forever a thing of the past.

While exponential population growth can be good in the short term, sufficient resources are necessary to feed future generations. Even though these human foragers were cognitively well equipped, they still remained subject to the availability of prey and plants, especially as they depleted the resources in each newly explored zone. They needed to ensure their safety in order to survive. But how? By observing nature and sharing knowledge, these clever beings understood how to germinate seeds and managed to domesticate wheat. At last, they were able to avoid starving during the winter by storing grain. The fields sown by human hands attracted herds of herbivores, who approached despite their fear of humans, their hungry bellies making them run the risk. *H. sapiens* gradually abandoned their nomadic hunter-gatherer way of life and settled down in one place. But for many, working the land was grueling. Cultivating plants involved plowing soil, sowing seeds, and harvesting crops, all by the sweat of one's brow. Dogs were of no use in this work. *H. sapiens* needed a new partner. Nine thousand years BCE, one of our ancestors had the idea to domesticate a new animal, one that was admired for its power and had been in their vicinity since they had established their settlements: the aurochs, a now-extinct wild ox. Its physical strength and massive frame would have made it an invaluable helper in the fields, as well as a major source of food. It is likely that the same method used to domesticate the wolf was implemented once again: very young aurochs that were easier to handle were captured to accustom them to humans. The young aurochs became attached to humans, just like the wolf had. Our ancestors had planned a completely different destiny for this animal, however. They bred the aurochs, penned them in enclosures, forced them to work in harsh conditions, and finally ate them. Unlike the dog, who even slept in humans' beds, the role assigned to the aurochs was to help humans in the field and then later end up on their plates. It was a real success. Humans were able to cultivate even larger areas of land and thus harvest larger quantities of barley and wheat. This knowledge was transmitted between different populations and through generations. *H. sapiens'* population continued to climb.

Strengthened by this discovery, *H. sapiens* did not stop there. They domesticated other animals, such as mouflons (wild sheep), wild goats, and wild boars. Over generations, these animals became the infantile and docile versions of their ancestors: cows, sheep, domestic goats, and pigs. *H. sapiens* tried their hand at domesticating many other species with varying degrees of success. When they encountered wildcats ten thousand years ago, the experience was very different from how wolves or animals raised for food were domesticated. Wildcats prevented grain stores from being raided by rodents.[25] Later, around 4000 BCE, tribes set out to conquer an animal whose spirit and majesty have always fascinated humankind: the horse. The task was much more arduous than it had been with aurochs and wild boars. This freedom-loving animal disliked being penned up, even after being tamed. Horses had been expected to be field workers similar to aurochs, but these animals required perseverance, a yoke, and a whip to finally succumb. But humans glimpsed a quality in this animal they had not been able to harness until then. More than any other animal, the horse was built to run. In the adventurous brain of one of our ancestors sprung the incredible idea to climb astride. While riding a horse seems natural to us now, at that time no animal had ever before attempted to forcibly sit upon another to benefit from its speed. The audacity coupled with the imagination of *H. sapiens* is endless. This crazy invention symbolized human domination over animals and transformed the ancient world. Trade flourished. Humans could migrate much faster, conquer new territories by waging war in new ways, and export their knowledge with unparalleled speed.

Surprisingly, in the beginning, domesticating animals did not substantially change the way humans thought about their place in nature. Their beliefs were still strongly rooted in their animist heritage, which was founded on the premise that spirits inhabited animals. Five thousand years BCE, the Vedic religion in India held that there is a oneness in the substance that makes up the universe, which makes the body and mind inseparable. It taught that nothing distinguishes humans from animals—"souls" can be incarnate in one just as much as in the other. Some two thousand years later in ancient Egypt, humans worshipped each animal species and even went so far as to believe in an afterlife for animals. They mummified their pets so these animals could accompany them into the afterlife. But it did not stop at pets. From the ibis to the mongoose, the Nile perch to the scarab beetle, it is estimated that 70 million animals of various species were mummified during ancient Egyptian times. The embalmers used the same mummification techniques and refined substances for animals as they did for humans. Humans and animals were considered the incarnations of deities, each occupying a vital place in the world. Like electricity running through hundreds of different cables, the Egyptians believed that each living being was animated by the same divine creative energy. Animals were the earthly representations of the gods. They worshipped the bull Apis and the cat goddess Bastet, to name just two. Recent studies measuring the level of animal protein in the hair of Egyptian mummies show clear evidence of a diet very poor in meat and fish, similar to that of present-day lacto-ovo vegetarians.[26] The sanctification of animals drastically curbed their consumption. Over the centuries, however, zoomorphic representations were abandoned in favor of hybrid half-man, half-animal beings, like Thoth with the body of a man and the head of a sacred ibis, Anubis with the head of a jackal atop a human body, and Bastet, who became a woman with the head of a cat. This transition to depicting gods in the image of humans suggests a shift in thinking. At this time, the Egyptians felt the need to identify with their deities and began to idolize the divine substance that animates animals more than the animals themselves. This explains why many animals began to be raised for sacrificial purposes and their mummies were sold as offerings in the temples.

In the 5th century BCE, like the Egyptians, the Greeks worshipped several deities. But while Egyptians still venerated animals or gods with animal heads, beasts and men coexisting in the same symplasm,* the twelve gods of Olympus all had the distinctive characteristic of being anthropomorphic, without any bestial features, although there remained some traces of animist cults here and there. The Greeks glorified gods and goddesses who resembled them, feature for feature. Human-animal hybrids became monsters. While Apis, with the body of a man but the head of a bull, embodied the creator god for the Egyptians, for the Greeks, the Minotaur (a chimera born from the love affair between the human woman Pasiphaë and the Cretan Bull sent by Poseidon) was a being that fed on

---

* The "symplasm" is an intracellular mass formed by plant cells. The cytoplasms of the connected cells form a single compartment shared by all the cells, just as humans and animals were perceived as being part of a whole.

human flesh, had uncontrollable impulses, and needed to be locked up in a labyrinth. Theseus's fight to kill the hybrid beast reflects the hero's victory over his own animal nature.[27] Another terrorizing creature, the sphinx, with the body of a winged lion and the head of a woman, was known for ravaging fields. The centaurs, horses with a human bust, were less frightening, but they symbolized the bestial appetites—lust and drunkenness in particular—that one must learn to control. Far from the tender portrait of famous storyteller Hans Christian Andersen's Little Mermaid (an undine), the sirens of antiquity were winged enchantresses (without a fishtail) who used their song and music to seduce navigators, luring them close in order to devour them. In ancient Greece, animals thus continued to nurture spiritual life, but their mystical attributions substantially altered the way in which they were perceived in the real world. In this anthropocentric belief system, nonhumans only had a utilitarian function and were not offered an afterlife. To move closer to the refined and the celestial, we had to deeply bury the *beast* within us, without ever letting it escape. This is how humanity thought it could rise, by shedding its animal status. Animal sacrifices in honor of the gods were commonplace, and the techniques used in domestication were rife with cruelty.

Only a few great thinkers, such as Pythagoras, Plutarch, and Ovid, opposed the treatment inflicted on animals and opted instead for a strict vegetarian diet. As Plutarch said,

> You ask me upon what grounds Pythagoras abstained from feeding on the flesh of animals. I, for my part, marvel of what sort of feeling, mind, or reason, that man was possessed who was the first to pollute his mouth with gore, and to allow his lips to touch the flesh of a murdered being: who spread his table with the mangled forms of dead bodies, and claimed as his daily food what were but now beings endowed with movement, with perception, and with

voice. How could his eyes endure the spectacle of the flayed and dismembered limbs? How could his sense of smell endure the horrid *effluvium*?[28]

All the other schools of philosophy (Aristotelianism, Epicureanism, Platonism, Stoicism) regarded humans as the only recipients of divine substance. This differentiated them from animals, which had been denied language and reason. Aristotle stated, "Plants exist for the sake of animals and the other animals for the good of man. . . . If therefore nature makes nothing without purpose or in vain, it follows that nature has made all the animals for the sake of men."[29] This utilitarian vision, revisited by later religions and great theorists, had a considerable impact on Western civilization and has managed to persist to the present day.

While the Roman religion borrowed the majority of its gods from the Greeks, it did modify them somewhat and also created its own myths, linked to its own foundation and history. The religion became increasingly centered around the imperial family, the deities making the emperors belong to the divine. In the Roman Empire, amphitheaters were the stage for great games in honor of the god Jupiter and brought together thousands of spectators. In the year 186 BCE, to celebrate the victory of Marcus Fulvius Nobilior over the Aetolians and the Asian king Antiochos III, an event was organized in Rome that would be remembered millennia later as the first great *venatio* (meaning "hunting" in Latin). Panthers and lions were starved for days and kept in the dark, then they were thrown into the arena for spectators to watch devour one another. Emotion ran high: people reveled in the bloody spectacle, enthralled by every bite, every tearing of a limb, every scream. In reality, something much more intimate than it seemed was taking place in the heart of the arena. By cathartic effect, the fascination with the morbid made it possible to ward off one's own anxieties about death.

Sitting safely in the stands, with no risk of being devoured, spectators rubbed shoulders with danger to feel more alive. The amphitheater was a stage upon which people could reenact the theater of the natural world. They could artificially recreate combat or hunting scenes using wild animals, while enjoying the luxury of being an observer with an exhilarating front-row seat rather than an active participant. The deadly spectacle was an unconscious reminder of their ancestral fear of the beast, both exciting and repulsive at once. Sharing in this communal thrill, people had the illusion of control, of omnipotence, and it reassured them of their place in the universe.

Some citizens, including Roman philosophers, expressed their disapproval, but the collective enthusiasm for these *venationes* was such that politicians used them to win votes, proposing increasingly varied twists to the sport, such as throwing men into the arena to go head-to-head with the beasts. In 55 BCE, Pompey staged over-the-top mock hunts during which hundreds of wild and rare animals died for the simple pleasure of "entertainment."[30] Pliny the Elder, in his *Natural History*,[31] described how twenty elephants were attacked with javelins by men condemned to death and knocked over the fence meant to protect the first rows of spectators at the Circus Maximus. He recounts,

> One of these animals fought in a most astonishing manner; being pierced through the feet, it dragged itself on its knees towards the troop, and seizing their bucklers, tossed them aloft into the air: and as they came to the ground they greatly amused the spectators, for they whirled round and round in the air, just as if they had been thrown up with a certain degree of skill, and not by the frantic fury of a wild beast. Another very wonderful circumstance happened; an elephant was killed by a single blow. The weapon pierced the animal below the eye, and entered the vital part of the head. The elephants attempted, too, by their united efforts, to

break down the enclosure, not without great confusion among the people who surrounded the iron gratings.

Under the reign of Caesar, no fewer than thirty-five hundred pachyderms were massacred. There was certainly no lack of imagination when it came to the choice of species to make *play* in the arena: crocodiles, hippopotamuses, leopards, lions, bears, and tigers all succumbed in great number to the sound of the audience's cheers. People's desire to bury their own animality ended up making it reappear in its most vile form. Under the pretext of entertainment or art, they had found a skillful way of extracting themselves from morality.

It was in this context of our growing denial of our animality that a new religion was born in the West, one that profoundly changed humankind's relationship with the cosmos: Christianity. For Christians, multiple deities were replaced by a single god who no longer expressed himself in natural things, because he himself was the creator of a world that was totally subservient to him. While the polytheistic gods were a part of a whole, the Christian god stood outside of his own creation. By imagining an all-powerful god separate from nature, humanity had to completely reinvent itself. Animals and plants were no longer the manifestation of spirits but creations of God intended to serve humankind. As cultural anthropology specialist Thibault Isabel points out, "sacredness in the world is replaced by sacredness outside the world."[32] Although little is known about the speed and manner in which this dogma spread within the Roman Empire, research carried out by specialists has concluded that Christians constituted a minority of the population until the 3rd century CE.

But what motivated the pagans to abandon their deities for this new god? There was, of course, the promise of eternal life after death, while the Roman religion did not guarantee the afterlife.

Likewise, an innovative foundation imbued with fraternity gave hope to the most humble classes. But by breathing a divine spark into humans and not into animals and by making humans creatures in the image of God, the Christian religion offered above all the immeasurable feeling of being unique in God's creation and infinitely superior, whereas paganism was content to consider humans a mere piece of a larger puzzle. Emperor Constantine, plagued by many visions, was receptive to the appeal of this religion from an ideological point of view. But above all, according to Yves Modéran, a French historian specializing in Late Antiquity and the High Middle Ages, Constantine discerned, through his conversion to Christianity, the opportunity to "enhance his claim to supreme power through divine legitimacy . . . which would establish two things: an all-powerful god, who alone was superior to all the gods of pagan Rome, and a universal god of the whole Empire, who would transcend borders."[33] Under Constantine, the number of new converts grew exponentially, but he remained relatively tolerant of the followers of paganism. This was not true of his sons Constans, Emperor of the West, and Constantius II, Emperor of the East, whose desire to evangelize the whole world was evident. Finally, in 391 CE, Emperor Theodosius I made Christianity the official religion of the Roman Empire. Pagan temples were closed, and the great games, such as *venationes* and gladiator fights, were prohibited, because in addition to their bloody and bestial nature, Christians regarded them as the expression of "mysterious and evil forces associated with the archaic and cruel deities of the Roman Pantheon."[34]

From this point on, humanity no longer needed to bury its animal nature to give itself substance since its break with nature allowed it to expel itself from the kingdom of beasts. The concept of the animal became associated with impurity and imperfection, and anything that evoked instinct or carnality took on a strongly negative connotation. In medieval Christianity, many animals were demonized to the extreme, to the point that they were often attributed evil intentions. Cats, crows, and toads, to name but a few, were associated with witchcraft and thought of as incarnations of Satan. The Church saw devil-worshipping heretics everywhere. Incidentally, in the Apocalypse, the devil will take the form of a beast. Animals that most closely resembled humans became absolutely repulsive to them. For instance, the upright posture shared with bears, the anatomical similarity in common with pigs, and the way apes imitate human behavior were all seen as diabolical and disgusting.

From the 13th to the 17th century, animals were attributed the moral capacity for their actions, because paradoxically their demonization made them conscious beings. This *humanization*, however, was granted to them only when made to answer for their bad deeds. In 1120 the bishop of Laon, France, pronounced the excommunication of caterpillars and field mice for having ravaged the crops. At that time, when public life and theology blended together, it was common to drag an animal before a tribunal for judgment. The French medieval historian Michel Pastoureau recounts the tragic fate of the sow from Falaise in his book entitled *Une histoire symbolique du Moyen Âge occidental* (A symbolic history of the Western Middle Ages).[35] It was in Normandy in the year 1386 CE that this pig was brought before justice for having fatally injured an infant by biting its face and body. The pig even had a lawyer representing it. Despite the fact that pigs had been prohibited from roaming the streets freely since the death of Prince Philip of France in 1131 because a piglet darting out in front of him had caused him to fall from his horse, it was still common to see swine amble down the streets and in houses two centuries later as they helped get rid of household waste and other filth. This Falaise sow attacked an infant when passing by its cradle, biting the baby all over its body. The pig's owner was not fined, but as punishment, the human-eating beast was dressed in human clothes and white gloves, then mutilated in the same places on its body as the baby had been, and finally hanged, like a human would have been. Some historians have concluded that at that time, animals were perceived as humans. Having studied human-animal relations for a long time, I do not see it the same way. In my opinion, this way of humanizing animals by dressing them up or *offering* them a trial had nothing to do with the possibility that they might have been regarded as human.

In the Middle Ages, animals were not perceived to be sentient beings but rather creatures devoid of reason that were entirely submissive to humans. A particularly horrible event (like a baby being eaten alive by a sow) had to have a diabolical explanation. To kill a child, the beast must have been possessed by a demonic force and therefore by Satan himself. And how exactly was Satan depicted in medieval times? As part-human, part-animal! By making the accused animal look human, as with the sow, we sought to give it a devilish appearance, no more, no less. Doing so gave ourselves permission to condemn the animal. Éric Baratay, a French specialist of the history of human-animal relationships, gives another example of the way in which animals or plants could be judged indiscriminately. "According to the legend, which was taken very seriously at the time, Saint Bernard saw swarms of flies arrive at his monastery one day; it made him angry, so he excommunicated the flies, and the flies died. In the Gospels, Jesus curses the barren fig tree, and the tree suddenly dies. Clearly, if we can curse an animal or a plant, thereby sending it back to the devil, that means we can judge its faults."[36]

Later during the Renaissance, various philosophers invited us to reevaluate humankind's position in the universe. Michel de Montaigne discusses human vanity in "An Apology for Raymond Sebond," which appears in his *Essays*,[37] emphasizing,

Can anything be imagined to be so ridiculous that this miserable and wretched creature, who is not so much as master of himself, but subject to the injuries of all things, should call himself master and emperor of the world, of which he has not power to know the least part, much less to command it? And this privilege which he attributes to himself, of being the only creature in this grand fabric that has the understanding to distinguish its beauty and its parts, the only one who can return thanks to the architect, and keep account of the revenues and disbursements of the world; who, I wonder, sealed for him this privilege?

Montaigne grants animals a morality that should serve as a model for humans to restore their connection with nature. This great man developed many revolutionary ideas for the time, including an animal's ability to reason and feel things, but his anthropomorphic vision of animals was often criticized.

While God was central in medieval thought, the Renaissance placed humans at the center. The humanists remained believers, however, and the animal condition hardly changed. In his letter to the Marquis of Newcastle dated November 26, 1646, the philosopher and mathematician René Descartes wrote that even while he conceded that animals were sentient, he did not consider them to be anything more than mechanisms, whose absence of language and thought differentiated them from human beings. His famous phrase *cogito, ergo sum* ("I think, therefore I am") was the foundation of Cartesianism and presupposed self-consciousness in order to be a living subject (as opposed to a mechanical object), a faculty animals lacked according to him. He reduced animals to an assemblage of parts and cogs capable of feeling pain through physical mechanisms but incapable of thought. His mechanistic vision had a profound influence on Western society, and yet it was criticized by other great thinkers when *Discourse on the Method* was published.

About two centuries later, the Enlightenment proposed to sur-pass obscurantism and depicted all the important issues of the time using images of animals. The philosopher François-Marie Arouet, known as Voltaire, was outraged by the fate of animals and did not believe it was the will of God to consume meat. He wrote, "We re-gard this horror, often pestilent, as a blessing from the Lord and we have prayers in which we thank him for these murders. What can be more abominable than to feed constantly on corpses?"[38] For Jean-Jacques Rousseau, having spiritual attributes or the capacity for reason was not a requirement for animals to have rights;[39] and since they were able to experience suffering, that in itself was suf-ficient to condemn acts of cruelty committed against them. The German philosopher Immanuel Kant took up these ethics centered on the sentience of animals not only to denounce their mistreatment but also to prevent the moral degradation of humans. He saw that "he who is cruel to animals becomes hard also in his dealings with men."[40] In his *Critique of Judgement*,[41] Kant opposed the animal-machine theory. According to him, because humans are different from all other beings, we have only indirect ethical duties to ani-mals because they are not rational beings and therefore do not have rights, or duties to other beings themselves.

The French Revolution, whose revolutionaries carried on the Enlightenment's ideals, marked another shift in the way in which humans and animals were regarded. Just like humans, animals were among the king's oppressed subjects, and this led to a more sensitive consideration of their plight, much like during the Enlightenment. This outlook was also the result of a deep anti-religious sentiment, which the *sans-culotte* François Boissel echoed, paving the way for a new conception of humanity. In his book *Le Catéchisme du genre humain* (Catechism of the human race),[42] he emphasized that "all religions are nothing but inventions of man, transmitted and per-petuated by the ignorance and credulity of the majority." His work had a considerable influence, putting an end to the Church's anthropocentric beliefs and inviting a reevaluation of the animal race. Livestock shows were banned in the streets of Paris in 1793. A year later the French National Museum of Natural History was created along with the Ménagerie, the zoo of the Jardin des Plantes, whose objective was to study animal behavior and which housed the abandoned animals from the royal zoo of the Palace of Versailles. Another eight years passed before anyone began questioning the mistreatment of animals. At the end of the 18th century, Parisians no longer wanted to see cows and sheep being clobbered and butchered on the banks of the Seine. This led to the creation of enclosed slaughterhouses, which concealed the distasteful sight from the view of passersby.

Since its creation in 1804, however, the Napoleonic Code (French civil code) has considered animals to be personal property, despite numerous public discussions with citizens on the matter. In the 19th century, Victor Hugo used his eternal talent and pen in his tireless fight against slavery and social misery. As an activist for the right to education and the rights of women and animals, Hugo chaired the Anti-Vivisection League in 1883 and advocated vegetarianism as a new cornerstone of humanity, stating, "As long as man feeds on animal flesh and torments animals, something wild will remain in him and he will know neither health nor peace." For Hugo, the way we treat animals not only determines the fate of humanity but also our ability to master what is most cruel and vile in our own animal nature. His poems beseech us and evoke how closely linked the destinies of humans and animals are. Hugo writes, "Who knows how their fate is intertwined with our fate? . . . Who knows if the misfortune we cause to animals and if the useless servitude of animals does not end with Nero on our heads?"[43] Under pressure from this great thinker and from the general and politician Jacques Delmas de Grammont, the French National Assembly adopted the first law devoted to animal protection in July 1850. Nine years later, Darwin's book *On the Origin of Species* was published, in which he pulverized the anthropocentric world of Christianity with his theory of evolution. He dismantled our relationship with the cosmos with a scientific argument, not philosophical one. Humans' simian nature shattered the myth of our supremacy over all living things. It forced us to reconsider our animality and to reconstruct the essence of humanity.

Despite the accumulating biological evidence of an evolutionary continuum between humans and animals and the many initiatives in favor of a further reflection on the animal condition, *H. sapiens'* exponential population growth during the 19th and 20th centuries created a race to increase production, casting aside any ethical considerations. By 1700 there were approximately 700 million human beings. This number reached 2 billion in 1930, and by 1999 the population had tripled from there. To feed all these mouths, we had to produce even more, and we had to innovate. Through science, new methods were developed that intensified vegetable farming. Modern farming techniques were so effective that we could afford to feed and raise more animals. This "progress" in livestock farming led to a dizzying increase in the population density of animals. Confined, pregnant, and butchered at an increasingly young age, animals were valued only for their slaughter weight and were merged into collective and utilitarian names, such as "cattle" and "poultry." In this competitive race, the price of meat fell drastically between the 1950s and 1980s, which caused meat consumption per capita to soar. The post-rural urban world became completely disconnected from that of animals, which remained out of sight.

Totally separated from animals and nature, humans today (and Westerners in particular) claim their supreme place in the universe while no longer knowing how to justify it. How do we conceptualize our relationship with animals when we do not even physically see some of them anymore, like the cattle that nourish us? The fact remains that while we can endlessly debate this philosophical question, science has unveiled things in recent decades that we cannot ignore. And what do these discoveries show us? In light of this new ethological knowledge, are humans so different from animals?

2

Intelligence of Their Own?

Man had always assumed that he was more intelligent than dolphins because he had achieved so much—the wheel, New York, wars and so on—whilst all the dolphins had ever done was muck about in the water having a good time. But conversely, the dolphins had always believed that they were far more intelligent than man—for precisely the same reasons.

DOUGLAS ADAMS,
SO LONG, AND THANKS FOR ALL THE FISH, THE HITCHHIKER'S GUIDE TO THE GALAXY, 1984

A few decades ago, anyone who was overly interested in animal intelligence was considered a romantic. Today some people legitimize their dominant place in the universe by their degree of intelligence, which they consider superior to that of *beasts*. When we use this term, we mean either "any animate beings other than humans" or "stupid people who have little to no intellect." Which is to say that the term has taken on a pejorative connotation. For many of us, a "beast" calls to mind the image of a cognitive pyramid, with the base made up of lower beings, ascending to higher and higher cognitive levels, and finally reaching humans at the very top. The more an animal is genetically close to us (and therefore resembles us), the more we are inclined to concede a certain level of competence to it.

Conversely, if an animal is morphologically very different from us, we are less inclined to attribute any intelligence to it at all.

For example, we readily admit that chimpanzees have some form of intelligence, but we ascribe hardly any intelligence to mollusks. In fact, our empathy toward a particular species decreases the longer ago our evolutionary divergence with it was.[44] We find it more difficult to adopt prosocial behavior toward animals that are very different from us. Our intuition leads us to believe a lot of other things to be true about intelligence, such as the assumptions we make related to physical size. The smaller the animal, the less intelligent we consider it to be. Insects, for example, are often dismissed as being mechanical beings with purely instinctual behavior. The environment in which an animal evolves also tends to influence how we perceive its intelligence. For instance, our perception of fish and other aquatic creatures like octopuses is that they only obey their impulses. Dolphins and whales are exceptions to this since we know they are genetically closer to us.

In reality, recent discoveries reveal that none of these beliefs has any scientific basis. They are the result of our anthropocentric perception of the world. While there are multiple definitions of "intelligence," the scientific consensus is that this concept encompasses all the skills related to the ability to adapt to a situation. Adaptability has several components, including the capacity for rational thought, learning, comprehension, and problem solving. It is through these components that scientists evaluate intelligence. When I collaborated with the Clever Dog Lab in Austria to assess the cognitive performance of dogs, this specialized team created several behavioral tests structured around key concepts, such as memory (quantified by the ability to retain learning after several months) and social cognition (assessed by the ability to follow various human cues, such as finger pointing).[45] Whenever ethologists study the cognitive performance of animals, they try to design tests specifically adapted to the species they are studying. This endeavor is so difficult, however, that even the most seasoned among us regularly fail and sometimes have to start from scratch after years of work. Scientists are also humans who perceive their environment through their own sensory system, which has its own limits. Even when trying to overcome our natural tendency toward anthropomorphism, we still struggle to capture the infinite richness and plurality of worlds in which other species evolve.

Each Species Has Its Own Image of the World

According to David Hume, a Scottish Enlightenment philosopher, if nothing forces us to believe in an objective reality, we must nevertheless act as if that reality exists. The perception of reality is not exact because it passes through an intermediary: us. The different information we receive using our five senses allows us to build a mental representation of the world. Our sense of sight, for example, results from the reception of light signals (called "photons") traveling to the cornea and the lens, which then deflect the light to focus it on the retina. And here in the retina is where an incredible phenomenon occurs: Cells within the retina emit nerve impulses to the brain depending on the signals they receive. The *raw* data that travels to our brain is compressed, however, much like an ultra-pixelated photograph. There are even gaps in this image, due to blind spots between the retina of each eye and the nerves that relay information to the brain. The brain then works some magic to build a single image by adding in the signals coming from the two eyes. This fills in the gaps of the pixelated *photo* to render a three-dimensional impression. Therefore, how we see the world is truly *created* by our brain. Since each animal species has its own brain and its own sensory system, there are as many mental representations of the world as there are different brains. Given these countless perceptions of the world, we have to wonder if it could well be that certain species mentally represent the cosmos in a way that is more faithful to reality than we humans do.

Consequently, as ethologists, when we study a behavior, we must observe an animal by projecting ourselves into its world. We must learn to disregard our own senses and to cast doubt on our own mental representations. The German pioneer of ethology, Jakob von Uexküll, elaborates upon the concept of *Umwelt* in his books *Streifzüge durch die Umwelten von Tieren und Menschen* and *Bedeutungslehre* (published in English as a single volume titled *A Foray into the Worlds of Animals and Humans*),[46] explaining that even when organisms share an identical environment, they experience different worlds. Where humans see only a flower, bees perceive a hypnotic spiral shaped by ultraviolet waves that direct them precisely to the nectar. When dogs walk alongside humans, they are not simply wandering down the sidewalk, but rather they are immersed in constellations of scents that they are hastening to sort and memorize. While the human eye is only able to intercept light rays in a single direction, a dragonfly's multitude of three-dimensional eyes provide a panoramic view and an almost unlimited depth of field. Humans often think of our nights as being silent, but in reality they are punctuated by symphonies of ultrasound composed by nocturnal animals, most of whose vocalizations are imperceptible to us. These different ways of being alive remind us how relative our perception of physical reality is.

Given these animal mental worlds, animals should no longer be regarded as inferior beings but as subjects in their own right. How then should humankind define itself?

We Are Not Born Human,
We Become Human

H. sapiens have come to control the world through the construction of collective knowledge. At the individual level, this aptitude has been accompanied by a growing cerebral capacity over the course of our evolution. Does this mean that what we define as specific to humans and our capabilities, which at first glance seem superior to those of animals (such as language), should be functional from birth? In other words, are we born "human," or do we become "human"?

Around the 5th century BCE, Herodotus recounted the story of a pharaoh who, wishing to know the first language of humanity, entrusted two babies to a shepherd to raise them among his goats. History does not tell us the outcome of this experiment, though we suspect a fatal one. In the 13th century, Emperor Frederick II of the House of Hohenstaufen, renowned for his avant-gardism and being a man of immense culture—in fact he was nicknamed *stupor mundi* ("wonder of the world") by his contemporaries—also sought to reveal the primordial language, but he was never able to decide if it was Arabic, Greek, Hebrew, or Latin. In 1211, several newborns were entrusted to nurses with strict orders to care only for the infants' vital needs and to never speak to or even interact with them. The Franciscan friar Salimbene di Adam described the cruel end of this story, writing, "It was for nothing, because all the children died. . . . Indeed, they could not survive without smiling faces, caresses, and loving words of their nurses."[47] Around the year 1500, King James IV of Scotland, also known for his erudition, repeated the "forbidden experiment" (as the American writer Roger Shattuck later called it),[48] placing two infants in the care of a deaf woman who could not speak to them and whom he left on a desert island to keep all three away from any human influence. Again, the children perished.

At the beginning of the 19th century, the idea of raising a baby far from its mother and deprived of family became ethically unacceptable, but the question of an original human nature remained unresolved. Many scientists were interested in cases of feral children because, growing up outside of human society, their cognitive development could provide keys to understanding human nature. Psychiatrist Philippe Pinel studied the case of a feral child from Aveyron in France, who was found at the age of twelve. He had been living alone and unclothed in the woods near Lacaune, feeding on acorns and roots. Pinel considered the child to be mentally ill, abandoned after being abused by his parents, as the burn marks on his body and the scar on his throat seemed to show.[49] When Jean Itard, a physician at the French National Institute of Deaf-Mutes (which later became the National Institute for Deaf Youth in Paris), took charge of the young boy's case, making it well known in history with the publication of *Mémoire sur les premiers développements de Victor de l'Aveyron* (published in English as *An Historical Account of the Discovery and Education of a Savage Man*),[50] he had a completely different opinion of this child, however. A pioneer in child psychiatry, Itard was convinced that the abnormal development of this boy, whom he named Victor, could only be the consequence of his isolation. According to Itard, a person only becomes human by learning from peers; we are not innately human. Itard began his book by stating,

> Cast on this globe, without physical powers, and without innate ideas; unable by himself to obey the constitutional laws of his organization, which call him to the first rank in the system of being; Man can find only in the bosom of society the eminent station

that was destined for him in nature, and would be, without the aid of civilization, one of the most feeble and least intelligent of animals. . . . In the savage horde the most vagabond, as well as in the most civilized nations of Europe, man is only what he is made to be by his external circumstances.

Despite working with Victor for four years, Itard only made slight progress. He was able to teach his protégé to communicate some emotion, however, and to do some work. Victor had great difficulty speaking and died at the age of forty, without being autopsied. Some two centuries later, François Truffaut brought this touching story to the screen with his film *L'enfant sauvage* (*The Wild Child*).* In other reported cases of children growing up without any social connection, they all presented cognitive deficits and an absence of language. When pediatricians and ethologists investigated these young people's pasts, however, they observed that, like Victor, these children had very complicated family histories, with their abandonment being the culmination of abuse suffered since birth.

It is therefore difficult to determine to what extent life in the wild impacted their mental development. In my opinion, the more recent case of Oxana Oleksandrivna Malaya allows us to lift the veil and glimpse our original nature. This young girl was born in 1983 in the village of Nova Blagovishchenka in Ukraine. Oxana was abandoned at the age of three in the backyard of her home and grew up among a group of dogs for almost five years, until she was reported and taken to social services. When she was found, she was crawling on all fours and behaving like the canines with whom she had lived. She drank water by lapping it up and showed her tongue in quick back-and-forth movements, much like panting. The sounds she

---

* *L'enfant sauvage* (*The Wild Child*) is a 1970 film directed by François Truffaut with Jean-Pierre Cargol in the role of Victor de l'Aveyron and Truffaut playing the role of Dr. Itard.

made were nothing like the ones a human makes when imitating a dog. On the contrary, they sounded unmistakably like the intimidating growls produced by canines. The videos of Oxana, recorded a few years later, made news around the world and raised a wave of protest. Many viewers were offended by the teenager's behavior, most likely because it brought back the specter of our animality. Scientists did have a significant piece of information about Oxana, though: when she had been examined at birth, she had shown no signs of abnormal development. Having carefully viewed the video recordings of this child (who had been dressed in clothes for the filming), I believe them to be authentic, particularly because Oxana moved on all fours with such dexterity and expressed such a diverse range of canine behaviors. Even an astute expert in dog ethology would not be able to imitate these descendants of the wolf so well.

In the absence of a human role model, Oxana had used the animal she lived with as a model for learning. She had grown up with dogs for five years and exhibited all the behavioral characteristics of these animals, including how she communicated, which did not include a single human word. Unlike Victor of Aveyron, Oxana managed to learn to speak again after being taken into care, but not without difficulty. In her misfortune, she had been abandoned at the age of three (which is after the critical period of language acquisition that the American linguist Noam Chomsky defined as being from birth to three years old) and had been introduced to human speech as a baby, which increased the likelihood that she could relearn language and her species' behaviors. Still, without access to a base of knowledge and without a human role model, our animality has free rein, revealing a *blank* original nature, devoid of culture and beliefs.

In fact, various studies have established that the absence of socialization induces cognitive delays in all young social mammals, and humans are no exception. In the 1960s, the American psychologist Harry F. Harlow demonstrated that depriving new-born rhesus macaques of their mothers and all other social stimulation resulted in major deficits in their mental development.[51] But unlike these macaques, Oxana did not grow up alone. In fact, it was quite the opposite, seeing as she had lived among dogs. On closer inspection, the young girl had shown tremendous adaptability by managing to survive with these animals, which testifies to her remarkable behavioral plasticity.

Childhood is a delicate period during which the brain absorbs an extraordinary amount of information. Each time it receives an environmental stimulus, new neural connections are established. Deprived of *normal* stimuli, feral children still use their capacity for learning, but their role models are the animals with which they have learned to live. Therefore, they do not master human language and instead communicate through cries or sounds that have a similar tone, frequency, and motivation to those of their nonhuman learning models. Most feral children move on all fours and exhibit behaviors identical to their animal models. What troubles humans, who have become social and cultured beings, is that relieved of their human *finery*, they are *nothing more than* animals after all. Millennia of beliefs and cultures have propelled humankind light-years away from our original environment, so much so that we have forgotten that we belong to the organic world.

Is There an Essence of Humanity?

In the West, the concept of "humanity" only formed in opposition to the concept of animality. Greek philosophy considered intelligence to be the criterion for defining humankind and the divine. Christianity reinforced this by considerably widening the divide between humans and beasts. And, to finish convincing us of humanity's total supremacy, some philosophers have proposed a utilitarian view of animals, giving us permission to exploit them for our own use with complete peace of mind. As we have seen, however, ontologically speaking, nothing distinguishes a human being from other animals. Without cultural transmission, we behave like animals, at least that is if we are lucky enough to grow up among them.

Our cognitive abilities—initially limited to meet our physical needs—reached new levels only with education. From this transmission of knowledge came the manifestations of human genius. Nevertheless, the only way this common base of knowledge in humans could develop was through skill acquisition over time. It is precisely these skills that the champions of human supremacy use to justify the enslavement of other species on Earth. But what skills are we talking about exactly?

ARE TOOLS PROOF OF OUR SUPREMACY?

The use of tools has long been held up as clear evidence of our exceptional intelligence. This ability requires intentionality, meaning that an individual must be able to imagine a specific objective in order to use a tool. It also indicates advanced cognitive abilities. Many see the mastery of tools as a major cognitive leap, foundational to our evolutionary success. As a matter of fact, intelligence was necessary to have the idea to strike two stones together to make a sharp tool that could be used to butcher carrion and save the skins. Until a few years ago, this highly distinctive trait was associated with *Homo habilis*, "handy man," who appeared 2.6 million years ago in Africa. It has long been believed that a more arid environment would have favored bipedalism and therefore liberated the hands, thus allowing the handling of tools. But in 2012, archaeological digs led by the French prehistorian Sonia Harmand of Stony Brook University in New York State prompted us to reconsider the idea that tool mastery is an explanatory factor in the origins of humanity.[52] In Kenya, west of Lake Turkana, Harmand's team of archaeologists found tools at the 3.3 million-year-old site of Lomekwi, which predated the appearance of *H. habilis* by five hundred thousand years and somewhat undermined the theory that the use of tools was the founding pillar of the human race. Also discovered at the Lomekwi site were fossil remains of *Kenyanthropus platyops* ("flat-faced human from Kenya"). This species was a tree-dwelling, bipedal hominid from the Pliocene, and therefore distant in time from *H. habilis*, the biped whose hands were thought to have been *freed* since they were no longer needed for gripping tree branches.

Since that discovery, ethologists have demonstrated that a wide variety of species use tools, knocking out the final bricks in the theory that this skill is a specifically human cognitive act. Scientific literature is overflowing with examples of tool use, so I decided to share a few that show great creativity. For instance, chimpanzees are masters in this field, using various tools. Their tool-use behavior is transmitted intergenerationally to such an extent that we acknowledge genuine chimpanzee culture.[53] Some of these primates lightly chew leaves in order to use them as sponges because, when soaked, the leaf-sponges make it easier to drink liquids (such as water[54] or even palm wine, an alcoholic beverage people produce from the raffia palm that is collected in containers in the chimpanzee's habitat[55]). Other times, meticulously chosen sticks are plunged into the bottom of anthills to remove the tasty insects, a selection process that becomes more refined as chimps age.[56] When attacking termite mounds, chimpanzees use a combination of two tools and adjust their behavior according to how the mounds are configured. For underground termite mounds, the chimpanzee first uses a rigid stick to perforate the structure, then uses a thin stem to serve as a fishing rod. For external dome-shaped termite mounds, the chimpanzee opens the holes in the galleries with a sturdy branch before inserting a thinner stem that is frayed at one end and coated in saliva, which makes the termites stick to it so they can be pulled out from their nest. In Côte d'Ivoire[57] and Guinea,[58] chimpanzees are able to crack palm nuts starting at an early age. They position the nuts on a hollow support, then smash them using a tool whose material (stone or wood) and shape are carefully selected according to the hardness of the nut. But the inventiveness of this species does not stop there. In fact, a video captured in the forest of Bakoun in Guinea shows chimpanzees fishing for seaweed by using long sticks that they immerse in the water and then slowly bring to the surface in order to harvest the precious aquatic plants.[59] In Fongoli, Senegal, chimpanzees create actual weapons to go hunting.[60] They select a branch, remove its leaves, and then chew the tip to *sharpen* it. When they embark on the hunt for prey, they go in search of tree hollows, then thrust their spears inside in the hope of catching small nocturnal primates, called galagos, which sleep there during the day. If the spear comes out empty-handed, the chimpanzees sniff it, probably checking for the scent of their prey.

As for orangutans, they build remarkable nests for sleeping that take the mechanical properties of wood into account.[61] These structures have a solid scaffolding made of rigid branches that are skillfully woven together as well as a soft and comfortable mattress made from thin branches. Elephants, renowned for the impressive size of their cerebral cortex, are also capable of modifying branches, which they hold in their trunk and use as fly swatters.[62] While in captivity at the National Zoo in Washington, DC, an elephant named Kandula had the clever idea one day to use a cube as a stepstool, thereby accessing food more easily. This revealed that pachyderms have "insights" (akin to the famous "Eureka!" moment). Subsequently, Kandula applied this learning to any object likely to serve as a tool.[63] The wild boar, however, has the opposite reputation of the elephant and is often regarded as being crude. At the Ménagerie of the Jardin des Plantes in Paris, a Visayan warty pig, which usually lives in the forests of the Philippines, was observed grabbing a piece of bark in its mouth. The pig used the bark like a shovel to dig in the ground and, by adding other elements too, prepared a cozy nest for its young![64]

Other species too, such as ones that fly, never cease to captivate scientists with their ingenuity. Crows are proof of this fascination, having been attributed legendary powers since the dawn of time. Believed to be intermediaries between the world of the gods and that of humans, these animals were considered by the ancient Greeks to be disobedient messengers, egotistical, and bearers of bad news. Centuries later, Christianity associated crows with paganism; their hoarse cries, necrophagia, and black plumage proof of their evil affiliations. These beliefs became so rooted in superstitions that these birds are still considered to be "bad omens" today. To bring science back into the debate, and to shed light on the mysterious behavior of these corvids, Christian Rutz and his colleagues from the department of zoology at the University of Oxford equipped New Caledonian crows, known for their ability to use tools, with miniature cameras to track their movement.[65] The video recordings showed that these animals make picks to extract beetle larvae hidden in cavities of the candlenut tree. One crow carried a stick over a distance of more than 100 meters, (over 100 yards) demonstrating crows' ability to remember where they left their best tools so that they can use them again later. Far from their diabolical image owing to their plumage, these eminently social animals have a range of tools whose manufacture and configurations are as complex as they are numerous.[66] New Caledonian crows are even capable of innovation. They can work together to combine several sticks with no outright function, like pieces of a toy building set, to make compound tools such as poles.[67]

Debunking the "bird brain" cliché, a young Goffin's cockatoo named Figaro was observed having an "Aha!" moment when he used a stick in an innovative way to retrieve the nut he wanted from behind a metal grate.[68] The prodigy then learned to *carve* his tools, discarding the pieces that were too short and making those that were too long smaller. The story might have ended there with this incredible discovery, but the researchers wanted to know more and used Figaro as a model for other Goffin's cockatoos to learn. Of the twelve parrots tested, six watched Figaro demonstrate his achievements, while the other six watched the tool move by itself (using a magnet) or observed a scene where Figaro received a reward without having to use a tool directly. Some of the individuals who witnessed Figaro's prowess were able to use tools in turn to access their own food reward.

Let's now immerse ourselves in the marine world, where some species also demonstrate surprising cognitive feats. In California, sea otters smash shellfish against rocks to feast on the animals inside, or even lay stones directly on their chests and use them to crack open mussels and other shellfish.[69] The frequency with which these carnivores use their tools depends on the availability of prey. The rarer the shellfish are and the more complicated they are to open, the more time and energy the sea otters exert in using their tools.[70] In Shark Bay in Western Australia, many dolphins search the ocean floor for food. They do not use their rostrum (the elongated part of their head), however, which could be injured by coral. Instead they use already torn sponges, allowing them to search the bottom of the ocean in order to feed on sandperch and avoid hurting themselves.[71] Dolphins learn to use this sponge tool by observing their mothers.[72] These cetaceans also use the empty shells of giant gastropods as fish traps. They entice their prey to enter a shell, then they insert their rostrum while lifting the shell to the surface of the water before shaking it to make the captive fish fall directly into their mouth. In this case, learning is happening horizontally, meaning that cetaceans learn this hunting technique by directly imitating their peers.[73] In Louisiana, some crocodiles (which we often picture as beasts devoid of mental life) position twigs and sticks on their heads, attracting birds in search of materials for building their nests. Tricked into approaching, the birds are devoured by the reptiles. The crocodiles not only use lures, but they also time this baiting strategy and use it more frequently during the birds' nesting season.[74]

Among invertebrates, the remarkably intelligent octopus, which increasingly fascinates the scientific community, can work together with other octopuses to assemble coconut shells that they use for protection.[75] While biologists once thought that these solitary cephalopods only met to copulate, the discovery of two sites several meters in length in Jervis Bay off the east coast of Australia, respectively called Octopolis and Octlantis (a contraction of Octopus and Atlantis!), turned our knowledge upside down.[76] Octopuses build walls and shelters, particularly with the help of clamshells that are meticulously stored for the occasion. These *cities* serve as meeting points where cephalopods gather, exchange information, and chase off foreign individuals. We find it hard to imagine the incredible cognitive faculties of these animals because when we consider them, we intuitively apply two erroneous beliefs: first, that the relationship between intelligence and genetic proximity to humans is proportionate; and second, that the more different an animal is from us or the more different the environment in which it evolves is from ours, the less developed its intelligence is. Cephalopods have long been victims of our blind spots. These skeleton-less animals with an extraterrestrial morphology are nevertheless endowed with 500 million neurons distributed between their brain, optic lobes, and suction-cup covered *arms*. These animals are here to remind us of the thousands of ways intelligence is expressed.

In spite of all these examples, some people continue to boast about humans' particular mastery of tools. Three essential characteristics define the use of tools in humans: planning skills, manual dexterity, and the ability to select what material to use. Without meeting all three of these conditions, an animal cannot be considered to have cognitive functions that are as developed as ours. In all the studies just presented, none of these requirements is missing. The only difference is that humans' manual dexterity is swapped out for the crow's beak, the elephant's trunk, the agility of the otter's paws or the dolphin's rostrum, the ultrasensitivity of the crocodile's snout, or the tactile intelligence of the octopus's arms. But do animals have the same degree of creativity as humans, who not only use a tool but also anticipate how the tool will be used when they make it? This final requirement is met by large primates capable of making *insect-fishing poles* with a complex design[77] and even by primates that make projectiles by anticipating the need for them.*[78] Elephants that adjust the length of the branches they use as fly swatters[79] also meet this requirement, as do the New Caledonian crows that bend stems to make *larva-fishing poles*.[80]

DO NONHUMANS HAVE LANGUAGE?

The second ontological argument recited over and over again by the defenders of human supremacy is about language. Our way of speaking is unique in the animal kingdom, but does language just naturally appear in humans? Would a child be able to master human speech without a model? As we have seen through the examples of feral children and deaf children without access to sign language, the answer is no. Human language requires long-term learning. Certainly, we do have the innate capacity to be attentive to the sounds produced by our peers. Researchers have demonstrated that human fetuses respond to their mother's voice,[81] particularly during the last

---

* An adult chimpanzee at Furuvik Zoo in Sweden was observed preparing rocks and other projectiles in advance to throw at visitors.

trimester of pregnancy, as illustrated by the activation of their cerebral cortex recorded by brain-imaging techniques.[82] We thought we had it all figured out when a team of British researchers from the University of Oxford identified a gene called FOXP2, which they believed played a considerable role in human language skills.[83] They found that in one London-based family, a mutation in their FOXP2 gene had resulted in serious grammatical and lexical difficulties. The gene is more active in the left brain of girls aged four to five years than in boys of the same age,[84] revealing a gender difference in human language acquisition. But it is not enough to simply have this gene in order to learn to speak. Beyond the genetic aspect, there are anatomical features that make speech possible. The morphology of our vocal tract, characterized by a longer oral cavity and a lower larynx than other primates, allows for an incredibly vast repertoire of sounds.[85] There is therefore an innateness in our ability to speak. A human baby deprived of language during the first few years of life, however, will suffer an often irreversible delay. Without learning following birth, the child will not understand the symphony of words and will not be able to produce them. Therefore, language does not appear automatically and cannot be thought of as the divine spark distinguishing humans from animals. In fact, it does not even seem that language has been unique to *H. sapiens* since Neanderthals not only possessed the anatomical characteristics necessary to master language[86] but also had the same variant of the FOXP2 gene as modern humans.[87]

Chimpanzees, which do not speak, share nearly 99% of our genome, but they have a version of the FOXP2 gene that differs from ours by two amino acids. This means that language could have taken root in *H. sapiens* much longer ago in our evolutionary history than previously thought. This is the opinion held by geneticist and neuroscientist Simon E. Fisher of the Max Planck Institute. According to him, new mechanisms were not required for language to emerge;

rather its appearance was the result of adjustments to genetic pathways that were already present in our ancestors who did not speak. French professor emeritus of cognitive psychology Jacques Vauclair and his former PhD student Adrien Meguerditchian suggest that if we were to travel back in time, we would see that the way our ancestors communicated (initially gestural) began to combine gestures and vocalizations around four to seven million years ago, which was when our common ancestor with chimpanzees was alive.[88] The ability to voluntarily exercise voice control was a huge advantage, facilitating the exchange of information over long distances and at night. This bimodal communication system became more sophisticated over time in hominids. But how can the striking level of linguistic complexity observed in *H. sapiens* be explained? In all likelihood, a major change occurred when our ancestors expanded their groups of a few primates to establish true human societies. The need to transmit increasingly rich information to a rising number of individuals grew. Words served as social glue, and language was a means of achieving a highly developed level of collaboration. Language's main biological purpose was to strengthen social ties, thereby increasing group size and competitiveness against rival species. Undoubtedly, our ancestors experienced other phases of rapid change, which made them the greatest linguistic composers in the world, in particular when they faced major challenges. In 1985 the American anthropologist Randall White at New York University put forward the idea that a phase of accelerated linguistic evolution took place thirty-five thousand years ago.[89] This date sticks out to us because it coincides with the period when *H. sapiens* and Neanderthals coexisted.

If spoken language is the result of an extended learning process, then can other kinds of language exist? A few decades ago, sign language was not considered to be a true language because it was thought that it required less-developed cognitive faculties than those used for speech. Not speaking meant less reliance on abstract thought. The analysis carried out by the American linguist William C. Stokoe (a hero to the deaf community) revolutionized the way we perceive sign language. He was able to show that, just like spoken language, sign language has a phonology, syntax, and morphology. The denigration this language had suffered for a long time testifies to a major cognitive bias. We view any manifestation of intelligence that deviates from the human "norm" as a pitiful peculiarity, with the "norm" regarded as the height of intelligence. This is especially so when that intelligence requires a different sensory system. We pass over the fact that spoken language too is naturally accompanied by intense gestures. Until Stokoe's work, we were convinced that speech—brandished as the ultimate symbol of our capacities for abstract ideas—was the requirement for defining a real language. Sign language proves to us that the hand, facial, and body movements of people, who for the most part have never heard the sound of a word, transcribe the mental dexterity involved in spoken language with the same eloquence. Contrary to widely held belief, sign language is not a universal language; in some geographical areas there are as many sign languages as there are different deaf communities.

In light of the latest ethological discoveries, can we consider the existence of language in nonhumans? We know that animals howl, growl, bray, quack, moo, and hiss, but is this a real language? Ethologists prefer to use the term "communication" to remind us that there are just as many modes of communication as there are kinds of animals on Earth. Above all, this distinction allows us to examine this topic through the lens of multiple sensory systems, such as the olfactory system, and not just those responsible for hearing and speech. When two animals communicate, they exchange information that allows them to regulate their interactions. This means they each assess the individual in front of them so they can adjust their behavior accordingly. In fact, vocalization is far from the only mode of animal communication.[90]

Nevertheless, let's consider acoustic communication so we can compare it to how *H. sapiens* communicate. Ethologists have unveiled the meaning of certain signals produced in the animal world. The well-known hyena cries, which resemble sarcastic human laughter, vary depending on the sex, age, and social position of the individual within the group.[91] Sea lion pups recognize their mother returning from hunting mainly by her voice, which is called her "individual vocal signature,"[92] just as penguins are able to find their mother or father by listening to only two-tenths of a second of their call,[93] or even how lambs from the age of just forty-eight hours can distinguish their mother by her vocalizations.[94] Dolphins, which also have their own vocal signature, are able to learn to imitate the specific whistle of a fellow dolphin so as to call out to them and get an answer back—a unique way of giving each other names![95] Fallow deer bucks' groans are different in each male[96] and provide information on the individual's strength and age. Even mother crocodiles adjust the care they give their young according to the high or low frequency of their calls, which tells the mother about the age and size of her offspring.[97] Even more surprisingly, the ocean is not the world of silence that Captain Cousteau poeticized it to be. In addition to the incredible acoustic abilities of cetaceans, fish also coo, rattle, and scream! Researchers' hydrophones have recorded a true symphony.

In humans, however, language is not limited to vocalizing sounds. We also produce linguistic units that provide information about our emotional state or our identity. Scientists have long believed that the fundamental difference between human language and animal communication lies in the ability to refer to external objects or events by using words, which the human recipient is able to perceive and, more importantly, to understand. In our lingo we call this "referential communication." Songbirds are particularly well documented in scientific literature as vocal animals par excellence. Field research has shown a high degree of variation in the sounds they produce, in terms of duration, shape, and repetition rate.[98] Their alarm calls in particular are modulated according to predator type and distance, which "receivers" respond to by adapting their anti-predator behaviors according to the signals they receive. Various field studies of vervet monkeys,[99] Diana monkeys,[100] and Campbell's mona monkeys[101] have confirmed the existence of referential communication in nonhumans. These primates emit series of sounds that are acoustically different according to what type of predator is approaching (such as an eagle or a leopard). Prairie dogs too alert their congeners of impending danger by encoding information into their alarm calls about the color of clothing worn by individual humans and even their shape![102] As for bonobos, who are inclined to sharing food, they also produce acoustically distinct calls depending on whether the food they find is high or low preference. Individuals listening to the playback of broadcast sequences spent more time looking for the food when the message transmitted meant "tasty food!" rather than when it announced a less delicious fruit.[103] Chimpanzees intentionally produce a quiet "hoo" that varies depending on the context to mean travel, rest, or alert.[104] Even the animals we think of as having the least intelligence, such as guinea fowl[105] and chickens,[106] produce and perceive alarm calls and attribute meaning to them. Therefore, we can see that most species have an oral vocabulary.

But it does not end there. Great apes, like humans, use their hands extensively to communicate. An international team of ethologists took an interest in the gestural communication of human children in their second year of life and compared it to that of chimpanzees living in Uganda. Of the fifty-two distinct gestures recorded in human children, forty-six were identical to those of these great apes.[107] Our offspring thus share 89% of their gestures with chimpanzees. There is one exception, though—pointing is 100% human, at least when we do not teach it to great apes! We can say that at the beginning of our life, our mode of communication is . . . apelike.

But let us return to the particularities of our own oral language. Is the incredible richness of languages spoken throughout the world distinctly human? Indeed, the Bible explains our linguistic diversity through the story of the Tower of Babel. The people of Babylon, united by the world's original language, decided to build a tower that would reach the heavens. God punished them for their arrogance by creating a multitude of different languages, so that the Babylonians would no longer understand one another and abandon their endeavor. They dispersed across the earth, thus founding new peoples. Today we know that the plural development of languages originated because of the environments in which groups of humans evolved. For example, the number of distinct consonants and how we arrange them into syllables correlates with rainfall, temperature, terrain, and the amount of forest cover.[108] Therefore, human protolanguages evolved differently depending on where a particular group of humans chose to settle. What we imagined to be the divine was actually shaped by the same laws of evolution as all our other abilities. Incidentally, ethologists have looked at whether groups of animals of the same species but living in different areas could, like humans, produce distinct dialects. Here again, the literature on the subject abounds. Prairie dogs,[109] Japanese macaques,[110] Campbell's mona monkeys,[111] birds,[112] killer whales,[113] elephant seals,[114] and pikas[115] all produce different vocalizations depending on where they live and the social group to which they belong.

Beyond being able to vocalize meaningful sounds that vary depending on where they live, are animals capable of organizing these units of sound into "sentences"? In other words, is it possible to demonstrate a syntax that gives functional meaning to the arrangement of animal sound units? Using hydrophones, researchers at the Karadag Nature Reserve on the Crimean Peninsula recorded the conversations of two dolphins.[116] According to these researchers, the pulsed sounds produced by Yasha and Yana (the two dolphins in the study) are similar to human language. The dolphin companions combined several sounds (each sound representing a phoneme, or the equivalent of a word) thus creating "sentences" of approximately five "words." In Campbell's mona monkeys, males give "krak" calls in the presence of leopards and "hok" calls in the presence of eagles, and they add an "oo" suffix to alter these alarm calls when the danger is a nonpredatory disturbance.[117] Their call sequences can range from two to forty calls, with varied order and rhythm, which then generate different messages.

Both dolphins and monkeys respect an omnipresent conversational rule found in human language: they do not interrupt their conversation partner and only produce sounds when it is their turn. As it turns out, this conversational characteristic is much more universal than previously thought.[118] And that is not all. Several mathematical rules observed in human language are also found in animal language. Zipf's law, which predicts an inverse relationship in human language between the length of words and the frequency with which they are used, has been widely documented in the vocal communication patterns of bats,[119] Formosan macaques,[120] and marmosets.[121] Another law, called Menzerath's law, postulates that "the larger the construct, the smaller the size of its constituents." It was thought to be specifically human but is found both in the vocal sequences of geladas[122] and in the gestural communication of chimpanzees.[123]

Therefore, animals communicate to exchange information by producing meaningful vocalizations, arranging those sounds, and respecting universal rules. All these similarities to human language should remind us that the acoustic signals used by each species are extremely varied and that there is an incredible diversity of living things. While there are as many nonhuman languages as there are different species, many of their similarities are linked to the phenomena of convergent evolution (when species that have taken different evolutionary paths exhibit identical behavior under the pressure of the same environmental constraints).

Could it be that, unlike humans, animals resort to these different languages in "autopilot" mode, without having to use mental representations? What if animals produced sounds only through uncontrolled mechanisms or learned these sounds through associative learning (when a stimulus elicits a response)? Since large primates are anatomically unable to produce phonetic sounds in a similar way to humans, primatologist Duane M. Rumbaugh had the ingenious idea in the 1970s to use lexigrams—geometric symbols representing linguistic units—to designate foods, colors, and objects. This is how he taught Yerkish (an artificial human language) to a very young chimpanzee named Lana. She managed to master the lexigrams with extreme precision.[124] Tested twenty years later, Lana was still able to recognize the old lexigrams that she had been taught when she was young.[125] Sometime later, Dr. Rumbaugh was joined by Dr. Sue Savage-Rumbaugh, and the pair examined the linguistic abilities of bonobos. Among them was Kanzi, who particularly attracted the attention of the media with his impressive skills. The work of the two ethologists was criticized, however, when they dared to interpret these large primates' behavior as indicative of symbolic communication.[126] Their response to these criticisms was that their protégés' ability to understand spoken words, identify the symbols corresponding to these words, comprehend what lexigrams were used for, use lexigrams even when referents were absent, and above all learn all this through observation and not mere associative conditioning left little doubt as to the symbolic nature of their communication.[127]

Other examples corroborate the existence of mental representations in animals during vocal communications. Among them is the fascinating prowess of the gorilla Koko, who learned sign language from the American ethologist Penny Paterson. According to Paterson, Koko was able to use no less than a thousand different signs, understood spoken English, and responded in sign language to questions asked orally. These findings created division in the scientific community. According to the skeptics, Koko did not really understand the meaning of her actions and had learned sign language through simple conditioning. Her signed responses must have been unconsciously prompted by her caretakers.

The Clever Hans effect was first documented in Germany at the beginning of the 20th century. Wilhelm von Osten, a former mathematics teacher, believed animals could be educated in the same way as humans. Owner of a horse named Hans, he devoted four years to teaching Hans to read, do math, and even answer yes or no to various questions. Hans would tap his hoof to express himself. Surprised by his stallion's abilities, which prodigiously included counting and reading, von Osten exhibited Hans's prowess to the public, and the horse, nicknamed Clever Hans, became famous throughout Europe. Several people took an interest and investigated these claims, concluding that there was no deception. German psychologists Carl Stumpf and Oskar Pfungst noticed, however, that by removing or hiding the person interrogating him, Hans could no longer provide a correct answer. What was happening was that the horse was interpreting the unconscious body language of his questioner and thus knew from these cues what answer to give. Disappointed, von Osten sold Hans to a jeweler. During World War I, Hans was mobilized for war and his life sadly ended on a battlefield. Yet during his life, Hans had demonstrated the remarkable skill of being able to detect postural changes that were imperceptible to the average person. Did Koko use the same *stratagem* as Hans? That is what some scientists think. But not all specialists share this opinion. Anecdotally, Paterson tells the story of the time the gorilla Koko, who had never been taught the word "ring," managed to combine the signs for "finger" and "bracelet" to form the word "finger-bracelet."

The question of nonhuman languages using mental representations remains a touchy subject. To help us see more clearly, let's take a look at field research carried out on other primates. Mother orangutans are able to stop their alarm calls when a predator arrives to avoid detection. But in addition to that, once the danger has passed, mothers produce this alert call again in front of their child, even though there is no longer any reason for concern, suggesting a desire to pass on to their offspring the information linked to the event that just happened.[128] This ability to *discuss* past situations presupposes an ability to mentally project oneself in time. Remembering past events, as well as where-and-when information, is a skill known as "episodic memory." It has been demonstrated in many mammals, including great apes,[129] rats,[130] miniature pigs,[131] dogs,[132] and cats (who also remember what and when).[133] The skill also appears in several birds (scrub jays,[134] magpies,[135] rufous hummingbirds[136]) and even cuttlefish![137] Like these other animals, orangutans recall past events, but in the aforementioned study, they transmit the "concept" of danger that they associate with that event by means of an acoustic signal after the fact! Therefore, nonhuman language can engage mental representation processes.

Another example of concept mastery in an animal located on an evolutionary branch far removed from our own is the gray parrot, specifically one named Alex—an acronym for Avian Language Experiment—who belonged to the American ethologist Irene M. Pepperberg. His death in 2007, reported by the *New York Times*, caused a stir. But what was so special about this bird to trigger such an emotional response? Unlike large primates and dolphins, Alex (whose cortex size incorrectly suggested not very advanced cognitive functions) managed to achieve the unthinkable: by speaking words, he demonstrated abstract thinking. Alex was not just mimicking, as some still believe. He was able to master the concepts "identical to" and "different from," and "bigger than" and "smaller than," which not only presupposes the capacity to categorize but also a mastery of concepts. He also did not need to be shown an object to specify its color. He used adjectives and pronouns (such as "I" and "you"). He was creative in his way of communicating—he could lie, give orders, and even had a sense of humor! Alex also mastered recursion, which in linguistics is defined as a set of rules inherent in grammar and long considered an essential property of human language. He understood phonemes and how to combine them, and did so at a level of sophistication comparable to that of a three-year-old human child.[138] His brain, with a configuration that is radically different from a mammal's brain, proved that different evolutionary paths can lead to comparable levels of intelligence.

Despite Alex's astonishing skills, some would still argue that we have never seen parrots engage in philosophical discussions together or change the world. How can this difference between *H. sapiens* and nonhumans be explained? Do animals not think or think very little? Proponents of human exceptionalism elevate language as a fundamental characteristic for generating thought. According to them, it is impossible to have a thought without a vocabulary! Yet when my brain wanders, I do not feel like I am using words. It is as if the *substance* of my thoughts is on a deeper level. Testifying to my non-verbal intellectual wanderings is the fact that I regularly search for words, which underscores the extent to which ideas precede their linguistic form. It seems to me that words are mostly helpful in structuring and developing my thoughts. From this angle, to me language appears to be more of a tool for navigating the mind and, of course, for externalizing it.

Before humans even knew how to speak, we were animals who thought. It is because of our preverbal ancestors' increasing need for cooperation that they developed their language in an innovative way, and not the other way around. In other words, thought preceded language, and as a new means of expression, language's most important role was as an extraordinary social glue that favored inventiveness and cultural enrichment. Dreams (those escapes from reality, as described by the French neuropsychiatrist Boris Cyrulnik and demonstrated by various experiments in warm-blooded animals) are perhaps the most allegorical expression of thought processes that exist in nonhumans. In fact, humans do not use words much during REM sleep, instead mixing emotions and images in their dreamworld. But then, what do animals dream about? Scientists have managed to unlock some of the secrets of rats' dreams. By recording their brain activity while foraging for food in a T-shaped maze, a research team found that those same brain areas were activated again once asleep, showing that the rodents seemed to replay the spatial trajectory that had led them to a reward.[139] Their memories scrolled by in their sleep state at a different speed than in real life. The experiment illustrates the time distortion phenomena at play when mammals dream, which explains why we sometimes have the strange feeling of having experienced something for longer than our dreams actually lasted. Rats, these poor unloved animals, are therefore able to remember past events, imagine future experiences, and dream of things they desire.

Animals think, dream, reason, and master different forms of language. But what is unique to modern humans is their ability to push their linguistic creation skills to the limit. Transcending time and space, humans have managed to make their words transport them through the infinite universe of fiction. This is certainly humans' greatest linguistic achievement. But what was it that caused the "thinking animal" to metamorphose into the "metaphysical animal"?

FROM THINKING ANIMAL TO METAPHYSICAL ANIMAL

In my view, the answer lies in the emergence of the need to believe in *another* dimension. This need appeared very early in the history of humanity, but it was not specific to *H. sapiens*. It was Neanderthals who two hundred and fifty thousand years ago wondered about the afterlife, as evidenced by their burial places filled with flowers— proof of their belief in the existence of the hereafter. The French prehistorian Henry de Lumley, former director of the French National Museum of Natural History, traced the first appearance of burial rites to an even older period about four hundred thousand years ago. This is the date of the first known funerary offering, a biface made of red quartzite called Excalibur, found on the site of Sima de los Huesos in Spain near the bones of about thirty *Homo heidelbergensis* individuals, who were predecessors of Neanderthals. Most likely it was the awareness of death, fully developed in the common ancestor of *H. sapiens* and the Neanderthal lineage, that gradually gave rise to the need to believe in invisible forces. Our distant ancestors, therefore, did not feel the immediate need to explain the world because first they needed to give meaning to death in order to ward off their existential fears. They replaced the absurdity and horror of their mortal condition with the promise of a passage to an intangible life. This belief was so deeply embedded in our ancestors that today, faced with the fear of losing a loved one or of dying, even the most atheistic among us find ourselves calling upon nonphysical entities to help overcome these hardships.

This need to explain our own finite nature and the loss of a loved one was resolved thanks to a new neural mechanism, mystical belief. Faced with unbearable loss, our ancestors became increasingly curious about natural phenomena, which led them to ask the question "Why?" and convinced them that something somewhere must be pulling the strings. At a time when science did not exist, imagining an invisible world that governed reality and explained all physical phenomena was an effective adaptive strategy. In the 1970s, the English primatologist Nicholas K. Humphrey noted a "surplus" of intelligence in great apes that was most likely unnecessary for meeting their vital needs. This cognitive surplus, which particularly developed with the increasing complexity of social life, would have led our ancestors to perceive everything around them through the lens of cognitive skills adapted to group living. In turn, this led them to interpret nonsocial environmental factors as social facts, governed by intentionality, and to imagine the reasons that explain these events.[140]

So, if we had to define the essence of humanity, do you think it lies in this key moment in our evolution when the first metaphysical thoughts sprang forth and our ancestors began to believe that *another* world existed? Even though they had no idea what was unfolding, this step opened the door to a world of *possibilities*. But, unlike other hominids, *H. sapiens'* intellectual capital increased tenfold because of what I call the "melting pot of brains" or, in other words, the sharing of knowledge. What distinguished them from Neanderthals and other species of humans was the ability to unite around common beliefs and to transmit them. The complexity of their language and the development of their capacities for abstract thought went hand in hand with their cooperation skills— *H. sapiens* had to name the fictitious entities in order to share them with as many others as possible. Words offered them an infinite number of possible symbolic combinations. For millennia, their animist beliefs, where spirits pass from the physical world into the incorporeal world, enabled them to think about Earth by rationalizing the forces of nature and to establish a moral framework and rules of conduct. In turn, subsequent religions gave structure to their societies. But the intellectual power of *H. sapiens*, augmented by their societal organization, also allowed them to free themselves from the forces of nature. By thinking beyond reality and by creating "collective brains," they were able to innovate like no animal before, protecting themselves from hazardous weather, the variability of available resources, and even certain diseases.

Our ancestors distinguished themselves by developing abstraction skills. But does this mean that before they uttered their first "Why?" they were unable to imagine an intangible world? To find out for sure, let's look at our closest cousins, the chimpanzees. In May 1993, observations carried out by the English primatologist Richard W. Wrangham of Harvard University undermined the theory that only humans have symbolic cognition, which is the ability to have a mental representation of an object, person, or event (a "signified" is represented by a "signifier," such as language or gesture). In human children, we call this the game of make-believe. In his study, Wrangham observed an eight-year-old male chimpanzee named Kakama express interest in a small log found in the forest, which he carried with him while searching for food and brought to the place where he usually slept.[141] Alongside his mother, who slept in her nest, Kakama began to stretch his limbs skyward while keeping the log in the air, a behavior the primatologist likened to the act of "playing airplane" observed in chimpanzee mothers who make their children *fly*. The only difference was that Kakama used a substitute object (not a baby) to which he obviously attributed *another* meaning. After playing with his log, Kakama built a tiny nest that was far too small to sleep in but whose dimensions matched his toy. He stored his favorite object there before falling asleep at its side, within arm's reach. Was this chimpanzee playing with a "doll"? It is difficult to make a conclusion on an isolated case. But after years of observation, the team saw that Kakama was not the only chimpanzee to display this type of behavior. On numerous occasions, children of this chimpanzee community were recorded carrying around sticks that had no discernible function. Young females were more likely than males to play with this favorite object.[142] Consequently, it is difficult not to agree with this theory, seeing as there is such a striking similarity to human play behavior! This would corroborate the theory of mind in large primates and prove the acquisition of symbolic cognition in the child chimpanzee, like in the young human.

Even more surprisingly, in 2016 a team of researchers from the Max Planck Institute in Germany identified four populations of West African chimpanzees that had a habit of accumulating stones in particular places.[143] At these sites, the primates threw their projectiles with force at a few previously selected tree trunks. The thrown stones accumulated in the cavities or at the roots of certain trees, creating piles of stones scattered here and there. This behavior was transmitted from generation to generation and has mainly been recorded in male chimpanzees, but females transporting their young have also been observed exhibiting the behavior. Unlike tools, the accumulations from stones being thrown against trees do not seem to have any use. Two explanations have been put forward to explain this phenomenon. The first hypothesis is that the males, who bang on the ground or make noise to impress their fellow chimpanzees, throw stones at specific tree trunks to produce louder sounds and establish their authority, a variant of one of their natural behaviors. The second hypothesis, which takes the participation of the few observed females into account, is that this cultural behavior emerged from *another* motivation. Piles of stones, called cairns, are well known to humans, who build them to use as physical landmarks and in funerary or religious contexts. Even more intriguingly, Indigenous peoples living in the same environment as the chimpanzee groups under observation also created stone burial mounds near *sacred* trees that functioned as shrines.[144]

Is it possible that chimpanzees had rituals with a symbolic function? This is the hypothesis put forward by the German team, as well as by the American researcher James B. Harrod of the Center for Research on the Origins of Art and Religion, who discerned the precursory signs of religion from these behaviors, with the necessary caveats.[145] Other observations in chimpanzees support the hypothesis that religion emerged as a by-product of increased social skills. In 1971 the famous English primatologist Jane Goodall observed certain groups of primates perform a "rain dance."[146] During a violent storm, individuals brandished branches in the air like weapons, violently destroyed them against tree trunks, and then ran while striking the ground and emitting deafening cries. Only male individuals exhibited this behavior, while females and the young witnessed the spectacle, a behavior that is reminiscent of the charge led by males in the face of a predator. But, unlike the defense reflex, the "rain dance" is a cultural behavior passed down through the generations and only appears in certain groups of chimpanzees. The incongruity of this dance is that these primates, faced with a stressful event, do not behave as if they are dealing with the physical forces of nature but rather with a being that is acting with intention. If you look closely, there is no direct adaptive advantage to this behavior, except for giving them the illusion of control! This sheds light on the origins of our own religious behavior.

What can we conclude from all these studies? That the emergence of the first "Why?" and the belief in *another* world could only emerge on fertile breeding ground? While the first humans used their potential for abstract thought to disrupt the status quo, they inherited it from their predecessors!

IS CONSCIOUSNESS UNIQUE TO HUMANS?

Proponents of a fundamental dichotomy between human and beast also invoke the idea that consciousness is specific to humans. But what does that mean? It is highly probable that consciousness is expressed differently depending on the species; and for that matter, we still have trouble agreeing on its definition. Because it is impossible to step into the mind of an octopus or a turtle, however, ethologists have proposed several tests to assess the existence of consciousness in animals. The most famous of these is the mirror test, developed in the 1970s by American psychologist Gordon G. Gallup. The researcher's claim was that an animal capable of recognizing its own reflection in a mirror, as humans can, is capable of distinguishing between itself and the external environment and, therefore, is self-aware. Great apes, elephants, dolphins, pigs, magpies, and manta rays were successful in this exercise while others were not. This suggests that only the aforementioned species have high-level cognitive functions. But that would be forgetting the essential fact that most species' individual recognition processes rely on many other signals, not just visual ones. Failing the mirror test is therefore not synonymous with an absence of self-awareness. If we invented an olfactory or acoustic "mirror" test based on the animal's sensory processing, we would probably be amazed by the animal's ability to recognize its "self."

Humans are not the only ones who have thought. It is omnipresent in mammals and birds, and probably even in certain invertebrates, such as octopuses. Descartes's famous *cogito, ergo sum* ("I think, therefore I am") therefore does not only apply to *H. sapiens*. Animals not only think, but they reason and demonstrate a wealth of inventiveness and cognitive prowess. Some even manage to surpass us, such as the chimpanzee Ayumu (the son of the famous chimpanzee Ai, a research study subject at Kyoto University known for her high-level cognitive performance), who was able to memorize the location of numbers 1 to 19 on a screen. When the numbers were covered by white squares, Ayumu managed in record time to click on the white squares in the correct ascending order at a speed never achieved by a human.[147]

The incredible working memory of chimpanzees probably existed in our ancestors and most likely diminished over the course of evolution to make room for other cognitive skills that accompanied *H. sapiens'* expanding social ties. But chimpanzees have not finished surprising us. Ethologists presented a virtual reality exercise, much like a video game, to adult chimpanzees and humans of different age groups, in which the apes and humans had to move through a maze as quickly as possible by seeking distance-reducing routes. Chimpanzees demonstrated tremendous spatial intelligence, far surpassing that of human children and approaching that of human adults.[148] In fact, these large primates are not the only ones to surpass humans in certain cognitive tasks. Faced with a problem that has two solutions (one learned, the other not learned but more efficient), all the capuchin monkeys and rhesus macaques tested did not hesitate to opt for the most efficient solution, while 61% of humans did not risk it, sticking with the familiar strategy they had learned. This study thus illustrates a less developed cognitive flexibility in humans.[149]

Now that we have demonstrated the intelligence of animals, we can ask our next question: are animals aware of their knowledge? Humans have the ability to know that they know and that they do not know. To assess whether animals are also capable of this, researchers developed the uncertainty control paradigm, which allows an animal to not answer a question if it is not sure that it knows the correct answer. By selecting the "escape" option during the study, the animal refrained from answering and still received a higher reward than if it had chosen a wrong answer, but a smaller reward than if it had chosen the right one. This metacognitive skill has been demonstrated in chimpanzees,[150] dolphins,[151] rhesus macaques,[152] orangutans,[153] and rats.[154]

But scientists have gone even further. Not only do humans have the capacity to know that they know or do not know, but they also have the capacity to reflect on their responses and thus answer the question: "Am I sure I answered correctly?" Ethologists developed another test, called the "confidence assessment paradigm," that allows animals to make retrospective judgments about the accuracy of their responses. In this test, subjects can no longer abstain from responding (there is no "escape" option) and must choose between the two offered answers, then rate their confidence. After giving an answer but before receiving the reward, each subject must choose between a "low-risk" icon and a "high-risk" icon. The first guarantees a small reward whether the answer is right or wrong, and the second will give a larger reward but only if the answer is correct. Thus, if the animal chooses the high-risk icon when its initial answer was incorrect, it will get nothing, so it is a bigger loss than choosing the low-risk icon. The underlying hypothesis of this test is that a living being capable of making a retrospective judgment on its own performance will favor the high-risk icon when sure of itself and the low-risk icon when doubting an answer's accuracy. Rhesus macaques,[155] crows,[156] pigeons, and bantams (dwarf hens)[157] successfully passed the test, demonstrating that they are capable not only of knowing they know but also of accurately estimating their level of knowledge.

Given that animals are aware of themselves and of their own knowledge, then could they also be aware of the fact that their congeners have knowledge? Different experiments have shown that chimpanzees,[158] ravens,[159] and scrub jays[160] are all adept at attributing mental states to another individual, a faculty that is called "theory of mind." Dogs and cats[161] also understand human beings' intentions through finger pointing. This is a specifically human gesture that we use when trying to indicate an object, by pointing to it with our index finger. Over the course of their domestication, our pets have acquired the ability to understand us. For animals' theory of mind to be similar in all respects to humans', however, some scientists believe there is an additional required component: the ability to understand false beliefs. In other words, animals must be able in their own way to have the thought "I know you're wrong!" Humans master this faculty by the age of four. Using an eye-tracking technique, an international team of ethologists demonstrated that bonobos, chimpanzees, and orangutans can think in this way.[162] These three great apes' comprehension of false beliefs suggests that this awareness also existed in our common ancestor.

As we can see, it is not a unique consciousness that distinguishes humans from other primates within the animal kingdom. In fact, the difference is humankind's ability to share psychological states with most other people. This skill allowed humans to attain "shared intentionality,"[163] a necessity for the increasing complexity and growth of social groups.

Not only do some animals understand false beliefs, but they are able to decide to deliberately introduce these false beliefs into the minds of their congeners or even mislead them about their own intentions. This willful distortion of the truth, better known as "lying," is so common in humans that we tend to underestimate the cognitive functions behind it. I am not talking here about small deceptions, which could be learned through conditioning, but rather about strategies that require complex thought. The American primatologist Sue Savage-Rumbaugh reported how the bonobo named Kanzi, after she offered him the key to his enclosure, pretended to have lost it. Once alone, Kanzi used the key to get out. The art of lying is not a characteristic unique to primates. In another experiment, researchers placed a raven in a compartment where it could hide its food in small caches, a behavior frequently observed in this bird. A second raven observed the caching from an adjacent compartment with a window, while a third raven could not observe the caching since its view was completely obscured by opaque curtains. When the other ravens were allowed into the main compartment, the storer raven tried to ward off the observer but was not concerned about the non-observer who had not seen anything.[164] Aware that it was being watched, the storer pretended to leave food in a given place, then took advantage of the time the observer raven took to search for the cache to hide it elsewhere!

Scientists now agree that all sentient animals (defined as being capable of feeling emotions and having desires) possess a high level of consciousness. The question ethologists ask is no longer "Do animals have consciousness?" but "What levels and contents of consciousness do they experience?"[165]

3

Our Bestial Emotions

With mankind some expressions, such as the bristling of the hair under the
influence of extreme terror, or the uncovering of the teeth under that of
furious rage, can hardly be understood, except on the belief that man once
existed in a much lower and animal-like condition.

CHARLES DARWIN,
THE EXPRESSION OF THE EMOTIONS IN MAN AND ANIMALS, 1872

Imagine the incredible confidence it took Darwin to dare put down on paper the idea of an evolutionary continuum between humans and animals. This was at a time when religion had established human omnipotence, and many philosophers supported that idea, believing that only humans were endowed with a range of heavenly gifts. Just a few decades earlier, students were taught that the earth had appeared in 4004 BCE, a date estimated with expert precision by theologians. Yet it was in this puppetlike world that the famous naturalist established his theory of natural selection in *On the Origin of Species* in 1859.[166] But Darwin did not stop there. About fifteen years later, he took up the pen again to write the superbly illustrated book *The Expression of the Emotions in Man and Animals*, which curiously is less remembered.[167]

Why did we set aside Darwin's discoveries about emotions? When the scientific tsunami of natural selection swept over the globe, swallowing with it the world created for humankind, the narcissistic wound related to our animal parentage was deep. Now over one hundred and sixty years later and even though we understand evolution to be a scientific fact, we still have a hard time thinking of ourselves as *mere* animals, and we do all we can whatever the cost to avoid turning back into dust at the end of our lives. Unlike philosophical theories, which we can debate endlessly, Darwinian theory (which today is a theory only in name) was proven by many experiments. It not only disrupted the life sciences but also our perception of the world, finally providing *the* key to the original mystery. In a desperate attempt to save ourselves from the fate of beasts, we continue to protect the last bastion of humanity: our soul. Like a burning flame animating the body, irrigating the mind, and transcending death, the soul is this intimate kingdom where passion, love, and memory intermingle. It is the impalpable place where our emotions are shaped and renewed. Granting beasts the privilege of an inner life would mean not only tearing down the last wall between human and beast, but losing the essence of our being.

The ethological studies carried out in recent decades are nevertheless unanimous: all mammals, from rodents to humans, experience "primary" emotions (surprise, fear, disgust, happiness, sadness, and anger), which are triggered by universal brain signals through a set of brain structures called the "limbic system." The similarity of the areas of the brain involved across species is such that the favored hypothesis is that feelings do not differ from one species to another. These animals all move to the tempo of their emotions.

In Search of Happiness

Hedonism upholds pleasure as the foundation of happiness. From time immemorial, humankind has yearned to find happiness. For the sophist Callicles (a character in Plato's dialogues), happiness entails "fulfilling all one's desires as they arise, without repressing them," making pleasure a central element of our existence. Like *H. sapiens*, all animals are driven by an unquenchable thirst for life. Still, there is a sizeable difference between "wanting to live" and "loving to live" that has to do with emotions and more specifically with joy.

Pet owners have little doubt that their cats and dogs are able to feel this emotion—their pets' cheerfulness lights up their daily lives. But, while scientific literature is chock-full of experiments on the perception of pain in mammals, showing that they all feel it, investigations on joy in animals are surprisingly rare. In an "operant learning" experiment, researchers at the University of Cambridge in England devised a cognitive exercise in which heifers were conditioned to press a panel in order to open a gate, which gave access to a food reward. A control group could directly access the reward without having to solve the problem introduced by the researchers. The cattle tasked with solving the puzzle exhibited higher heart rates and moved more animatedly toward the food, while those given direct access to the reward showed no such behavioral changes.[168] Is the reaction of the puzzle-solving cattle joy for having risen to the challenge? This is what the results indicated, even if the authors also noted that additional studies would have to be carried out given the small number of animals tested.

So do animals, like humans, seek to be happy?

PLAY ENLIVENS EXISTENCE

Hedonistic creatures par excellence, *H. sapiens* have long believed themselves to be one of a kind in this way. When ethologists began to examine the question of play in animals, the public was somewhat bewildered by the continuum observed between humans and nonhumans. In fact, the existence of play was demonstrated not only in mammals but also in species as varied as birds, reptiles, fish, and even some invertebrates. In their great diversity, these behaviors are organized into three categories: locomotor play (appearing from birth), object and predation play (running after a ball, for example), and social play (such as play fighting, hide-and-seek, initiating sex). Regardless of category, play is an intentional activity with the primary role of practicing future skills, as well as acquiring social data and physical data about one's environment. It is a central activity in all vertebrates, giving them the means to survive into adulthood. And the list of behaviors that it promotes is long. Play facilitates searching for food, learning different rules of conduct, fighting enemies, repelling predators, seducing and breeding, and even building nests.[169]

Humans attribute a quasi-religious dimension to play, however, because by transcending the mediocrity of everyday life, the mind can wander *somewhere else.* This *play for the sake of play* is where we draw the line between nonhumans and humans. If we step down from our pedestal for a moment, could it be that this characteristic we think is unique to our species is more universal than it seems? In other words, do animals play, like humans do, just for fun?

During a family getaway to Martinique, sailing off the coast of Saint-Pierre (the city's ruins blackened by volcanic ash are evidence of its painful history), a hundred dolphins that were either curious or amused approached our boat. Brushing the hull and sticking their heads up as if to greet us, the cetaceans offered us a remarkable acrobatic show as they leaped out of the water. The most stirring performance was given by a young dolphin that, while jumping as high as it could, twirled twice in midair before going underwater once again. It delighted us a good ten times more with this aerial feat. These acrobatics had no immediate biological function—the dolphins were having fun, just for fun! And what about their famous "bubble rings"? These underwater rings are created by a sudden movement of their head and are well documented by specialists. Each ring is a vortex generated by the tip of the dolphin's dorsal fin into which air is released from its blowhole. Bubble rings have no purpose, except for playing—just for fun! Some specialists even consider them to be a form of art. Dolphins are not the only ones to engage in this type of playful activity. Video footage showing the *playful* moments of Beltex lambs has made the rounds online. Climbing up to the top of a haystack and back down again, the lambs take part in an activity similar to the "king of the castle" game, which consists of remaining as long as possible on top of a structure while the other players try to make you fall off. Footage has also been captured of a crow sliding with the help of an improvised sled (the lid of a plastic container) on the slope of a snow-covered roof. Evidently the bird never gets tired of this jubilant feeling, since once its descent is complete, it returns to the top, sled in beak, to have another go. And what about mother chimpanzees? Not only do they have intense tickling sessions with their young, but they also play the "airplane game" by spinning their child and the "I'm going to eat you" game by pretending to devour the body parts of their little one.

To physiologically validate the "play for the sake of play" hypothesis, neuroscientist Annika Reinhold at Humboldt University of Berlin and her team recorded the brain activity of rats. These victims of their bad reputation mill around in the dark and often provoke revulsion and disgust because they contaminate food and carry disease. Reinhold set out to investigate the emotions these animals feel during games of hide-and-seek with their experimenter.[170] Each time they found their human partner, the rats were rewarded with tickling sessions. Humans and rodents took turns hiding. The results were telling: neurons were differentially activated in the prefrontal cortex of rodents depending on the events taking place during the game of hide-and-seek, in particular during the search for their play partner or while they were overjoyed in their hiding place. Rats also play, just for fun!

The fact is, for humans, anything can be an excuse to have fun: gambling, sports, video games, board games. *Homo ludens** holds the prize for the world's most playful being. How can our exceptional attraction to play be explained? When our ancestors shed their animal status, in freeing themselves from the environmental constraints of nature, they lost much of their original biological equipment. Once perfectly adapted to their environment just as a bonobo or an orangutan is to theirs today, humans gave birth to less and less *finite* beings, who, once adults, were unable to fend for themselves out alone in the wild. Their fur disappeared; their tooth growth became delayed and their canines smaller; their fontanels no longer closed after birth; their jaws became less massive; and their bodies and faces retained a youthful appearance. Morphologically speaking, our ancestors did not gain any advantages from these changes. In fact, they lost everything that had once made them a species perfectly adapted to their environment. In addition to losing these adaptive specializations, the metamorphosis was especially brutal because these strange naked apes saw their muscle mass melt like snow in the sun in comparison to that of other primates.[171] By withdrawing from nature and therefore from certain evolutionary pressures, *H. sapiens* found themselves immersed in what the Dutch anatomist Louis Bolk called "permanent fetal state,"[172] better known as "neoteny." From an anatomical point of view, neoteny is a regression, but from a cerebral point of view, it is an innovation. Because this incomplete biological state

---

* As named by the Dutch historian Johan Huizinga of Leiden University in his book on the social function of play, *Homo Ludens: Proeve ener Bepaling van het Spelelement der Cultuur* (Groningen: Wolters-Noordhoff, 1938).

allowed *H. sapiens* to acquire an unrivaled level of brain plasticity, they accumulated knowledge like no being ever before. Newborns required a longer period of mothering, which lengthened the period they needed to grow and, along with it, the duration of learning. For the German philosopher and essayist Peter Sloterdijk, newborns were "crippled by prematurity . . . and could only survive in the hothouses of culture." This is how the eternal child became more inclined to play and seek pleasure, by whatever means.

But *H. sapiens* are not the only neotenic species on Earth. By taking control over nature, humankind brought other animal species into their neotenic apparatus. As we have seen, wolves—the first animals to be domesticated—are the most emblematic neotenic species. Thirty thousand years ago, in the wake of human beings, some of these wild animals became unable to fend for themselves in nature. They remained pup-like, wagging their tails and barking. Selected for their aptitude for cooperation, these wolves acquired new cognitive skills and developed communication capacities that helped them interact with humans. Dogs had just entered the arena! These canids' playful bow is the best-known invitation to have fun. Its back arched, front legs outstretched, tail wagging, body half-crouched, the animal is ready to pounce and interact with its human companion or fellow dog. In the ranking of neotenic domesticated animals, dogs are the all-time champions. They excel in the art of play and continue to play in adulthood.

Some twenty thousand years later, some members of the feline genus that moved closer to human habitations also underwent a process of domestication, but in a very different way from dogs. Cats were marked by neoteny,[173] but morphologically speaking, it was only slight as their appearance remained close to their wild ancestor. From the behavioral point of view, however, the change was significant.

Once domesticated, adult felines kept a number of infantile behaviors, such as purring, meowing, and kneading* just as young
(not adult) wildcats do. Kittens in the bodies of felines, like *H. sapiens* they have become inexhaustible players. But play is not an ar-
tefact of domestication. Most non-neotenic species also play. By
extending the duration of childhood, domestication has only intensified a natural inclination.

THE SMILE, A WINDOW TO OUR EMOTIONS

While "words fail to express emotions," as Victor Hugo poeticized,
our facial or bodily expressions often reveal our state of mind. We
raise our eyebrows when we are surprised; our lips curl down when
we cry; we wrinkle our nose when we are disgusted. It was on the
basis of these observations that in 1862 the French doctor Guillaume-
Benjamin Duchenne became interested in human facial expressions as markers of emotion. This forerunner in neurology had the
idea to use electricity in his physiological experiments by stimulating
different muscle groups in the face to induce different expressions.
His results, published in his book *Mécanisme de la physionomie
humaine, ou Analyse électro-physiologique de l'expression des passions* (*The Mechanism of Human Facial Expression, or an Electro-
physiological Analysis of the Expression of the Emotions*),[174] revealed that a human smile resulted in an elevation of the corners of
the lips and was made possible by the contraction of the cheek muscles. But he added that, to distinguish a genuine smile from a fake
one, another body cue had to be observed: the formation of small

---

* Purring only occurs in the wildcat between a mother and her young, while domestic
cats frequently purr in adulthood. Adult wildcats do not meow, but adult domestic
cats meow to communicate with their human companion. Kneading is the action a
cat makes by alternating pushing its right and left front paws against something. It
is reminiscent of the behavior of "drawing" milk from the teats of its mother. Cats
will knead their human's legs or blankets. Only domestic cats knead in adulthood.

expression lines at the corners of the eyes. According to him, when a smile originates because of emotion, it activates not only the buccinator and cheek muscles but also the orbicularis oculi muscle, near our eyes. In homage to his work, the "Duchenne smile" refers to an expression of true enjoyment. We believe this way of expressing joy is unique in the world.

Newborn humans produce these *spontaneous* smiles while they sleep, predicting their feelings to come. A Japanese research team from Kyoto University found the same smiles in sleeping infant chimpanzees[175] and even in a more distant cousin, the Japanese macaque.[176]

These fascinating discoveries trace the evolutionary origin of the smile back thirty million years, the date of the common ancestor between our ancestors and Cercopithecidae.* According to comparative-psychology researcher Marina Davila Ross at the University of Portsmouth in England, this contraction of the zygomaticus muscle appears to promote communication within social species. The Dutch ethologist Frans de Waal explains that in chimpanzees smiles are used less frequently than in humans, and they have several functions. Most of the time, they are nervous grins that demonstrate the animal is intimidated, but they can also be used to ease tension or seduce a partner.[177] In macaques on the other hand, smiles are never associated with a positive emotion. Depending on the species, in adulthood smiles reflect an absolute sign of submission. In humans, the ultimate social animals, facial expressions likely diversified to promote communication when groups grew from a few individuals to actual human societies. Using facial recognition tools, Paul Ekman, professor emeritus of psychology at the University of California San Francisco, counted no less than

---

* Large family of primates including cercopithecine monkeys, baboons, and macaques.

eighteen different types of smiles, from polite grin to loving smile to ecstatic, overjoyed, "How-beautiful-life-is!" sort of smile.[178] As anthropologist Moira L. Smith of Indiana University Bloomington reminds us, however, despite being innate and universal, smiling is also modulated by our culture.[179]

I recall a dalmatian who, when excitedly greeting a familiar person, would wag her tail frantically but also bare all her teeth, down to the gums. This surprising behavior intrigued me, especially since the dog only did this in front of a human being. The few times I ever saw her growl, the way she bared her canines was different and was accompanied by unmistakable body language. By digging a little further, I realized that this case was not unique. The smiling behavior seemed to be a characteristic shared by certain dalmatians, as well as some individuals of other breeds. How can we account for this unusual behavior, which does not exist in the dog's ancestor, the wolf? The most plausible hypothesis is that in contact with humans, these animals developed the ability to imitate this human facial expression. In other words, they learned to smile to show us, using *our* communication codes, that they too experience happiness in our presence. This astonishing behavior is specific to certain particularly communicative individuals. This hypothesis is even more probable following an experiment conducted by Japanese researchers at Azabu University that highlighted the ability of dogs to distinguish smiling human faces from expressionless faces.[180] Furthermore, this aptitude is not limited to just the detection of positive emotions but also of anger, as shown by the French ethologist Bertrand L. Deputte and his student Antoine Doll,[181] as well as by the Clever Dog Lab team in Vienna.[182]

LAUGHTER IS TIMELESS

We have now seen that *H. sapiens* are not the only ones who smile, but humans do have another way of expressing their cheerfulness—

laughter. These rhythmic contractions of the diaphragm manifest themselves differently depending on the person. While some people's laughter resembles funny gasping noises, others' sounds like a chicken's clucks or a pig's snorts. Whatever their sound, these spasms are all driven by the same mechanism. They can be provoked in various situations, but one in particular—tickling—triggers them involuntarily. Until a few decades ago, the laughter that was thought to be specifically human appeared to experts in the field as "antitheti-cal to natural selection," as Steven Légaré, a professor of anthro-pology in Montreal, pointed out.[183] English researcher Marina Da-vila Ross and her team from the University of Portsmouth made a major discovery in this area when they compared the tickle-induced vocalizations of juvenile bonobos, chimpanzees, orang-utans, and a siamang (a type of gibbon) with those of human children.[184] The results were enlightening. The areas sensitive to tickling were similar between humans and primates. Moreover, the audio recording of the vocalizations showed shared sound charac-teristics. Based on the congruence of the acoustic data, the research-ers were able to reconstruct the great apes' phylogenetic trees. A video captured at a conservation center in Guinea convincingly showed the incredible similarities of human and primate laughter. In the video, a young female chimpanzee named N'Dama was be-ing tickled by a trainer and was laughing uncontrollably. She tried, as human children do, to push the arms of the man tickling her away.[185] For the American professor of psychology Robert Provine, this reaction to tickling, which only happens when the tickles are from someone other than oneself,* plays an important role in dis-criminating self from nonself and in constructing our identity.[186]

---

* We do not laugh when we tickle ourselves because our nervous system cancels the effects when the stimuli come from our own body. This is fortunate because, as Robert Provine points out, otherwise we would be constantly tickling ourselves by accident.

Since laughter can be provoked reflexively by stimulating certain areas of the body, then could it also exist in animals that are less closely related to us? In 1999 Jaak Panksepp and Jeffrey Burgdorf of Bowling Green State University in Ohio published startling results. The researcher and his doctoral student's experiment involved tickling rats that were accustomed to humans and recording the sounds they made. The rodents emitted short, high-pitched chirping at a frequency of 50 kilohertz, which is inaudible to humans. The chirping sound was easily differentiable from the sounds rats make in other contexts.[187] As proof of the intense joy they felt, the rats asked for more tickles at the end of each test session. At the time, the claim that rats laugh was quite controversial, but the existence of "laughing" in rats was later investigated by other teams, whose results corroborated those of Panksepp. And in 2016 neuroscientists Shimpei Ishiyama and Michael Brecht at Humboldt University in Germany ended the debate by identifying a brain region where these candid giggles originate and which is common to all mammals.[188] When an animal is tickled, intense neuronal discharges occur in the somatosensory cortex. The same region is activated when rodents play together, which suggests that, like in primates, rats' "laughs" are an invitation to play and relax.

These studies inspired Patricia Simonet of the Spokane County Regional Animal Protection Service in Washington State who recorded the vocalizations of dogs with a parabolic microphone during play sessions. The happy animals emitted their easily recognizable breathy exhalations,[189] meaning that dogs too *laugh* in their own way. Beyond being a simple display of emotion, this behavior has several social functions in this species. When Simonet played back the audio recordings of the canids, not only did the sounds reduce stress in animals entering the shelter, but they also stimulated prosocial behaviors.[190]

Laughter existed long before *H. sapiens* walked the earth. The neural circuits laughter activates are located in the oldest parts of our brain. Young children are overcome by fits of laughter very early in life. Tickling is not the only cause of these involuntary, noisy exhalations; laughing also punctuates our conversations. Greg Bryant, a specialist in vocal communication and social behavior at the University of California, Los Angeles, conducted a captivating experiment in which he had people around the world listen to two Americans laughing together. From this audio track alone, all participants were able to identify whether the two were friends or not. Therefore, laughter is more than an emotional expression; it is also a reliable indicator of the quality of a human relationship.[191] A real social lubricant, laughter most likely became more sophisticated among our ancestors to promote cooperation. It was a way to defuse conflicts and help establish social status. But in the absence of tickling, what exactly are we laughing at?

We laugh for many reasons! Sometimes we laugh out of nervousness, other times out of politeness, and more rarely out of aggression. But more often than not, we laugh at comic situations. The French philosopher Vladimir Jankélévitch saw humor as "a means for man to adapt to the irreversible, to make life lighter and more manageable." Is our sense of humor born from the desire to thwart the inevitability of death? I do not believe so. From the age of eighteen months, when they have not yet become aware of their eventual death, human babies respond to comic events and are even capable of finding humor in a situation.[192] So, could humor have appeared in the course of evolution before *H. sapiens* made their entrance?

High-level cognition is needed to appreciate an unusual situation because this requires being able to grasp the incongruous relationship to reality. Humorous individuals must be able to anticipate the comic reactions of others. While we cannot (yet) bring back our distant ancestors to unravel the mysteries of the appearance of humor, we can take a look at their other descendants, such as our direct cousins, the great apes. Francine P. Patterson, the American primatologist who studied the abilities of the gorilla Koko for years, recounted the story of how one day Koko tied Patterson's shoelaces together, then asked Patterson in sign language to chase her! The Dutch ethologist Frans de Waal described the pranks of Tara, a juvenile female from the colony of chimpanzees he was studying at the Emory National Primate Research Center in Atlanta. After discovering a dead rat, Tara dragged it by the tail, taking care not to touch its body, then placed the dead animal on the head of a fellow chimpanzee who was sleeping. When the chimpanzee woke up with a start, "loudly screaming and wildly shaking her body to get this ugly thing off of her," Tara picked up the dead rat's body to go and place it on another chimpanzee.[193]

By no means are we aware of all the forms of humor used by nonhumans, but we do acknowledge that humans have particularly developed their capacity for humor. Sarcasm, wit, satire, the list of inventive ways *H. sapiens* have devised to express their sense of humor goes on.

Are We the Only Ones Who Cry?

Sadness, Jean de La Fontaine said, "flies away [only] on the wings of time."[194] An emotion *H. sapiens* know well, sadness is most often expressed by crying and accompanied by lacrimal fluid, better known as tears. In our Western societies, we interpret tears differently depending on the age and sex of the person crying. Some interpret newborns' cries as being tantrums, but they forget that babies only express immediate needs at this age. In adulthood, our interpretation of tears completely depends (oddly) on sex. Seeing a woman sobbing is relatively acceptable because she is, in the words of the

French philosopher Olivia Gazalé, "essentialized as an irrational being." On the other hand, seeing a man crying seriously injures his virility, since "emotional repression" is the foundation of his masculinity. In her book *Le mythe de la virilité* (The myth of virility),[195] Gazalé explains that giving a gender to tears is a recent phenomenon, which began during the second half of the 19th century when new gender stereotypes became part of the education of young boys. Before male tears became a mark of weakness, quite the opposite, they embodied a form of nobility. In Homer's *Odyssey*, Telemachus and his father Ulysses are overcome by violent sobs when Telemachus realizes that the young man in front of him, transformed by the goddess of war, Athena, is none other than his father, who had disappeared years earlier.[196] Christianity also has portrayals of abundant tears. As Jesus approached Jerusalem and saw it ahead, he "*wept* over it, saying, 'If you had known, even you, especially in this your day, the things that make for your peace! But now they are hidden from your eyes.'" Julius Caesar himself "*wept* before his soldiers just after crossing the Rubicon." In the 17th and 18th centuries, tears were shed in public because they uplifted the soul. It was not until the 19th century that we began to feel some modesty. Ever since, our sobs have symbolized an uncontrolled emotion that is usually attributed to women.[197] We can see how significantly our culture impacts the way we express our sadness.

Why do we attach so much importance to our tears? We detest all our bodily secretions (feces, saliva, sweat) with one exception: tears. While the other secretions remind us of our bestial nature, do tears, which do not exist in animals, hoist us up to the heavens? In tales, myths, and parables, there are hundreds of stories from around the world about statues that cry. Some Christians have witnessed droplets emanating from the eyes of statues of the Virgin and Christ, and believed these tears to be a miracle. Tears of blood, water, and olive oil were considered to be fraudulent claims by the authorities of the Catholic Church, however. Regardless, this desire to see tears on the cheeks of inanimate objects speaks volumes about our desire to connect tears to the divine. In reality, animals also shed tears, but not emotional ones. The function of these tears is *basic*, serving to lubricate the eyes or expel a foreign body, and unrelated to any emotion. Does this mean animals do not feel grief?

We all can picture a completely devastated dog or the look on a cat's face at the end of its favorite human's life. There are so many stories of animals that will not even leave the cemetery where that person is buried. So what does science say? Humans, like animals, shed basal tears with each blink of an eye to lubricate the eyeballs, a feature that appeared in vertebrates three hundred and sixty million years ago. This type of lacrimal secretion, which protects our eyes from infections and brings oxygen and nutrients to corneal cells, originated in amphibians and has been preserved in mammals, birds, and reptiles.[198] These tears show no difference in their composition with those of animals. Brazilian researchers at the Federal University of Bahia have analyzed and compared human tears with those of a variety of other species[199] and found that their tear components were identical to those of humans but that their composition varied according to the environment in which the animal had evolved. In addition to these secretions, some amphibians have developed eyelids. Contrary to popular belief, the primary function of eyelids is not eye protection but to evenly spread tears on the surface of the eyes to keep them moist and prevent corneal desiccation. Now we know what blinking is for. In addition to basal tears, like other animals, humans also secrete reflex tears when the eyes are irritated (for example, when chopping an onion). But humans also have a third type: emotional tears. We produce these tears when we are sad, overcome with happiness, or because of our capacity for empathy. These droplets do have an evolutionary explanation: they appeared quite late in history, one hundred and fifty thousand years ago, when our social relationships became more complex. For some scientists, this very strange way we have of crying (about almost anything) could have facilitated mutual aid in ever-growing groups by making our emotions *visible*—a powerful "cry for help." For other scientists, our tears might have helped us to overcome the stress of extreme situations by releasing tension. In fact, when we experience intense emotions, our tears contain a compound that distinguishes them from reflex tears called leucine enkephalin, a natural painkiller that acts in the same way as morphine! It is probably a mixture of these two hypotheses that explains the origin of tears.

Does this mean that before our ancestors had the ability to cry they were not able to experience sadness? That is not the case. Our predecessors, who buried their dead long before *H. sapiens* were around, were familiar with this negative emotion. They just manifested it *differently*. They probably showed sadness much like today's chimpanzees, whose eyes remain dry but who isolate themselves or appear depressed when faced with a tragic event. In sheep[200] and pigs,[201] the administration of a serotonin inhibitor (which some call the happiness hormone, though that is not its only role) induces pessimistic behavior. Unpredictable and aversive events[202] or social isolation[203] lead to the same result. Like humans, animals can experience severe depression or a state of shutdown, which makes life seem bleak. Furthermore, sadness is not an emotion specific to higher vertebrates. Julian Pittman, professor of biology at Troy University in Alabama, demonstrated that zebrafish could also be prone to depressive episodes. They were such successful animal models of depression that Pittman has used them as models ever since for neurobehavioral and psychopharmacological research.[204] What then of the fate of goldfish living in their bowls? This is the most emblematic case of mistreatment due to our ignorance of the animal world. Fish for the most part are gregarious, can feel pain, and are eminently curious. They need adequate space, environments rich in stimulation, and to share the joys of existence with their fellow fish.

When we talk about sadness, we naturally think of the most tragic moments that punctuate our lives: the loss of a loved one. Do animals experience emotions similar to ours in such a situation? In the *American Journal of Primatology*, an international team of primatologists reported a case that has since become famous of a wild chimpanzee mother in Zambia whose sixteen-month-old baby had died.[205] The mother carried her child's body for more than a day, then on the second day laid it on the ground and returned to it many times, running her fingers over the face of her dead baby. On the third day, she took her baby to the group of chimpanzees she lived with who meticulously examined the dead body, holding its mouth open as if to make sure that no breath was coming out. This is not the first time that such reactions have been observed in our closest cousins. In 1992 a female chimpanzee named Jire carried the body of her child Jokro, who died at the age of two and a half, for more than twenty-seven days.[206] The bereaved mother cared for her son's remains, grooming him regularly, sleeping by him, and showing signs of stress each time they were separated. Other observed cases are equally distressing. Noel, a female chimpanzee born in the wild and living at the Chimfunshi Wildlife Orphanage in Zambia, remained close to the body of Thomas, a nine-year-old male she had taken in when he lost his mother four years earlier. One day, sitting near the corpse of the young chimpanzee, Noel selected a firm stem of grass and proceeded to meticulously clean Thomas's teeth.[207] What could this behavior mean? Dental grooming in chimpanzees had been observed in a study from the 1970s as being both a self- and social-grooming activity,[208] but never before had dental grooming of a dead chimpanzee been observed. While researchers remain cautious as to its interpretation, they have not ruled out the possibility that this chimpanzee had the desire to take care of the body of a deceased loved one. This sheds light on the evolutionary origins of our own mortuary behavior. The study also reports the behavior of a male chimpanzee, Pan, who was very close to the deceased. "Pan's frequent visits, the fact that he fiercely chased away a bold youth who attempted to move the corpse, and his behavior toward his dead friend are striking, interesting, and unusual." In this exceptional situation, Pan established his authority within the group, even though he was not one of the dominant males.

Hominids are not alone in mourning their dead. Off the coast of a Portuguese island, a team of researchers from the Madeira unit of MARE (Marine and Environmental Sciences Centre) observed adult dolphins trying to bring a dolphin calf who had died of natural causes back up to the surface. Sometimes only one individual (probably the mother) showed this behavior, and other times several individuals took part in the funeral procession.[209] It is impossible to talk about the distressed behaviors observed after the death of a loved one without mentioning elephants. The existence of the famous elephant "cemeteries" has never been scientifically proven, but various studies have concluded that elephants show evident distress at the loss of a loved one. The English zoologist Iain Douglas-Hamilton and his team recorded the behavior of several elephants in a reserve in Kenya when a matriarch was in the process of dying. A female named Grace worked desperately to lift the dying elephant to her feet, but the matriarch eventually collapsed. Grace's intense vocalizations left little doubt about the state of distress she was in.[210]

The universality of sadness is perplexing, especially since we do not really understand what purpose grief has from an evolutionary point of view. It is still a scientific mystery. Since losing a loved one is felt both physically and psychologically in humans and animals as a feeling of "emptiness," then would this mean animals are also consciously aware of death? Human beings perceive the tragic nature of their existence, an interlude between two eternities. For that reason, from a young age humans acquire the certainty that time flows in a single direction, from the past to the future. The notion of the "arrow of time," which seems intuitive to us, is in fact a complex concept, which children only master around the age of six. Before this age, not only do children not understand that their deceased loved one will never come back, but they also do not understand that they too, like all living beings, are heading toward their own inevitable end. According to what we currently know, there is nothing to confirm that animals (even the most intelligent ones) fully understand this abstract concept of irreversibility. They probably do not think about the fact that their time is fixed, with a beginning and an end. On the other hand, animals are driven by a strong desire to live, evidenced by their avoidance of pain, predators, and death. Likewise, when faced with the imminent prospect of death, they exhibit immense fear, which those of us who have survived a terrible accident or a serious illness have experienced too. I am even convinced that many animals are able to sense that they are *leaving* and to feel the extreme anxiety caused by this prospect. Many dog and cat lovers hold on to the poignant memory of their sick or elderly animals who, a few hours or days before their death, run away, hide, or show signs of hyperattachment, seeking out a beloved human as the ultimate comfort.

At the Heart of All Beauty Dwells Something Bestial

What emotion is related to beauty? Is it solely a human one? Plato placed beauty in a sphere beyond sensory perception and reason.[211] Millennia later, Pope Paul VI affirmed in his beautiful address during the artists' mass in the Sistine Chapel that "we need you. Our ministry needs your collaboration. It is because, as you know, Our ministry is to preach and to make accessible and understandable, even moving, the world of the spirit, of the invisible, of the ineffable, of God. And in this operation, . . . you are masters. It is your craft, your mission; your art is precisely that of understanding the treasures of the heaven of the spirit and to enrobe them with word, with colors, with form, with accessibility."[212] Christianity elevates the artist to the rank of celestial traveler, capable of transcribing the world's splendor. But is humankind, connoisseur of beauty, alone in having a sense of aesthetics?

A Swedish study conducted on chickens, by researcher Stefano Ghirlanda of the Group for Interdisciplinary Cultural Research at Stockholm University, has somewhat eroded the myth of a purely human sense of aesthetics. The scientist taught six of these domestic birds to express their visual preferences by pecking at a touch-sensitive computer screen.[213] He trained four roosters to react to images of female faces and, conversely, two hens to react to images of male faces. Then he presented them with new images tending toward exaggerated sex-typical traits that were relatively symmetrical. The birds systematically chose the most attractive faces, which were also the faces that human beings judged to be the most beautiful. While this single experiment does not tell us if chickens and humans process facial images in exactly the same way, in terms of the brain mechanism used for processing faces, it does conclude that chickens do have an aesthetic sense.

In fact, examples of aesthetic preferences in birds abound. Every year during the mating season in Australia, there is a courtship ritual that borders on being a work of art. The graceful great bowerbirds go to great lengths to create and decorate structures called "bowers," which are used only for attracting and mating with females. Male great bowerbirds first choose an area that they line with sticks. They then raise two parallel walls by planting sticks vertically in the ground, intertwining them, and then bending them to form a perfect arch. At one end is a small court they decorate with stones, bones, and shells, a collection of objects called "gesso." For Australian ornithologist John Endler of Deakin University, beyond their phenomenal building skills, bowerbirds also have a high aesthetic sense.[214] The erected stick tunnel allows the females to see the male, who calls and sings passionately on his gesso and who throws his collection of decorative objects with his beak. A sort of bull's-eye, this structure allows females outside to see what is happening inside, without having to enter. If the female is attracted, she passes through the arch and enters the court where mating occurs. But the bowerbirds' courtship ritual is even more fascinating than that. In his study of bowerbirds, Endler made a baffling discovery—the males arrange their decorative objects in a particular way so as to create an optical illusion.[215] The smallest elements are placed at the entrance to the arch, then the others are arranged in order of increasing size so that the largest ones are inside the court. This creates an illusion called forced perspective that makes the court look smaller and the male look bigger in contrast, giving the bird an advantage in mating. Indeed Endler noted that the better the forced perspective, the greater the chances that more females would give in to the male bowerbird's charms!

Another fascinating mating ritual is that of the blue-footed booby, a marine bird that lives on the Galápagos Islands, in Mexico, in the Gulf of California, and on some islands in Ecuador. To win over his sweetheart, the male engages in a *dance* showcasing his large shimmering blue webbed feet, while offering the female a gift. The more intense the color of the male's feet, the more likely the female is to be successfully wooed.[216] In fact, the intensity of the blue coloring indicates that the fish the boobies eat contain a high content of carotenoid pigment. This coloration reveals good nutrition—proof that an individual is in excellent health. The mating ritual continues with the two partners positioning themselves facing each other and continuing their foot movements, synchronizing their upward and downward movements. This is followed by the touching of beaks and bowing in which the head tilts forward and the wings spread majestically.

And what about birdsong? Clearly the purpose of birdsong is not to satisfy our ear for music, but rather birds sing to communicate with their fellow birds and even to seduce them. Male vocalizations are extremely sophisticated and can be used to attract females. Interestingly, and contrary to popular belief, individuals belonging to the same species do not hum the same melodies. A study of canaries carried out by a team from the Laboratory of Compared Ethology and Cognition at Paris Nanterre University showed that the more varied a male's vocal repertoire is, the greater his chances of mating are.[217] Females respond differentially to the sound structures of males' songs—a true aesthetic stimuli! Birds, therefore, have a taste for beauty that is mainly focused on sexual preferences. A link between aesthetics and sexuality is also found in whales, whose songs are even more creative during their mating season.[218]

As the French ethologist Michel Kreutzer, who conducted the study on canaries, stated, "For many evolutionists, the taste for beauty finds its origin in the attraction of sexual partners. Then, one can imagine that this taste was secondarily extended to other domains."[219] As to these other domains, human beings cultivated them ad infinitum. Thanks to humankind's imagination and capacity for abstract thought, a multitude of art forms emerged. Humans capture precious moments through the medium of painting; religiously sculpt thousands of different materials, modeling clay and earth with finesse; apply makeup to bodies and faces; create numerous kinds of ornaments in the form clothing and jewelry; perform architectural feats; and dance to the rhythm of different sounds. One of these art forms in particular—drawing—played a key role in the evolution of *H. sapiens*. An international team of archaeologists made a significant discovery in Blombos Cave in South Africa when they found a seventy-three-thousand-year-old stone shard measuring 4 centimeters (a little more than 1.5 inches) in length. This stone fragment had a crosshatched pattern on it that could only have been drawn by human hands using pieces of ocher[220]—humanity's first crayons! The mastery of drawing with crayon reached an artistic pinnacle thousands of years later when animal paintings and engravings began adorning the walls of caves. But drawing was not just for depicting things or for communicating with the otherworld.

Let's fast-forward a few thousand years to around 4000 BCE. At this time our ancestors began combining their representations of beings and things with sounds—the infancy of writing. We can see that this intellectual revolution ensued because of our artistic and symbolic aptitudes. This was revolutionary because with the birth of the alphabet, the exchange of information became easier and consequently increased. The same happened with the development of trade. Above all, the "integration of the language of men into the visible," as the French historian Anne-Marie Christin wrote so beautifully in her article "Les origines de l'écriture" (The origins of writing),[221] preserved information and made its transmission possible from generation to generation. This is how the "melting pot of brains" (the pooling of knowledge) experienced such rapid advancement, since it is enriched with the skills and knowledge obtained with each new generation of human beings. This *ad libitum* expansion of the collective memory enabled a new form of art to emerge: literature. *H. sapiens*, who had acquired metaphysical thought tens of thousands of years earlier, made a new cognitive shift, which was about to propel them from prehistory into history.

What about other forms of art? Music, for example, transports us and moves us. Charles Baudelaire wrote about music, saying, "I feel all the anguish within me arise of a ship in distress; the tempest, the rain, 'neath the lowering skies, my body caress."[222] All over the planet, Master KG's "Jerusalema" makes us want to dance; Beethoven's *Moonlight Sonata* makes us weep; and "Hotel California" stirs our souls. But are our musical sensibilities unique on Earth?

Videos of a sulphur-crested cockatoo named Snowball have made him an international star. We have no explanation for his achievements. This beautiful bird living with his owners in the United States was capable of dancing to the beat of "Girls Just Want to Have Fun" by Cyndi Lauper and "Everybody" by the Backstreet Boys, and raised its large crest to each fast-paced melody. Were his movements an anthropomorphic illusion? This is what American neurobiologist Aniruddh D. Patel and his team from the department of psychology at Tufts University in Massachusetts sought to find out. After presenting Snowball with several familiar songs, the researchers subjected him to a more complicated exercise by playing his favorite music but with different tempi.[223] Assuming purely mechanical movements, this tempo change would have no impact on the way the bird moved. But that was not the case. Snowball adapted his movements, slowing or speeding up to stay synchronized with the tempo. For skeptics who have doubts about Snowball's prowess, I recommend watching the videos available online.[224] At least fourteen dance moves are shown in the video, including swaying his head side to side, swinging his body in a counterclockwise circle, turning his head in a semicircle up high or down low, lifting a leg or lifting a leg in sync with a head movement, and most hilariously, headbanging, reminiscent of the intense head movements of rock band AC/DC. Snowball never danced to receive food rewards or to seduce a mate. His *choreography* seemed more like a social behavior, for interacting with his owners. This discovery is captivating for its similarities with human behavior. In fact, the scientists titled their paper published in the scientific journal *Current*

Biology "Spontaneity and diversity of movement to music are not uniquely human."[225] These bodily expressions have also been found in dogs, cats, and chimpanzees, but unlike with Snowball, it has not been possible to demonstrate that these animals' movements are synchronized with the rhythm of a song. But all of these observations converge in favor of the hypothesis that musical sensibility is universal in animals. The many benefits of music transcend interspecies barriers too. Stress in dogs,[226] chickens,[227] and fish[228] decreases just by listening to a soothing melody. Baby rats who had listened to Mozart in their mother's womb obtained better scores during cognitive exercises,[229] underscoring music's role in facilitating learning along with greater synaptic plasticity.

Nevertheless, humans seem to be hypersensitive to music. Depending on the melody, it provokes a wide range of emotions. To explain this phenomenon, researchers Anne Blood and Robert Zatorre of the Montreal Neurological Institute at McGill University used an imaging technique (called positron emission tomography) on human subjects and demonstrated that listening to music caused the activation of brain circuitry involved in pleasure and reward— similar to circuitry activated by food, sex, or drugs. When we see a parrot move in response to music, we might wonder if similar anatomical links exist in them, perhaps in a different configuration. But Blood and Zatorre underscore a human particularity, stating, "Perhaps as formation of anatomical and functional links between phylogenically older, survival-related brain systems and newer, more cognitive systems increased our general capacity to assign meaning to abstract stimuli, our capacity to derive pleasure from these stimuli also increased."[230] Thanks to a unique neural mechanism, humans are able to enjoy music in their own way and assign meaning to it.

While some people perceive music to be a simple recreational activity, psychology researcher Emmanuel Bigand at the University of Burgundy Franche-Comté and cognitive science researcher

Barbara Tillmann at the Lyon Neuroscience Research Center attribute a major role in the development of our cognitive skills to music (which they describe in their book *La symphonie neuronale* [The neural symphony][231]). Supremely neotenic, the human baby is born immature and requires constant attention. Auditory stimuli are valuable aids for a baby, who does not have very good vision, in its relationship to the outside world, offering reassurance whenever visual or tactile contact is broken. The brains of infants, which are capable of hearing and learning their mother's voice even before birth,[232] are responsive to lullabies. These songs help them regulate their emotions by calming them or reducing their stress,[233] and contribute to building the mother-child bond.[234] We are therefore born equipped with a musical brain. But that's not all: before grasping the symbolic content of the words and being able to say them, the intonations used by the mother help her little one to understand the emotional content of her speech[235] and to communicate with her. A team from the University of Würzburg in Germany has shown that the way adults speak to a newborn, stretching syllables and exaggerating prosodic contours in a singsong voice called "baby talk," stimulates infants' alertness and influences their cry melody.[236] The child is able to modulate its own sounds by integrating the information acquired when listening to its mother speak, in particular when she uses "baby talk." This has an incredible consequence—babies cry in different languages! A *musical* communication exists between mother and child long before the young human even utters its first words.

Music stimulates our cognitive abilities from the beginning of our existence and throughout our lives, improving our ability to concentrate and commit things to memory. *H. sapiens*' love of music can probably be explained by the evolutionary advantage it gave them by contributing to their increased cognitive abilities. But in my opinion, what was decisive for *H. sapiens* was not only that they increased their individual intelligence but that they were able to pool their knowledge through the "melting pot of brains." And this could only happen when *H. sapiens* profoundly changed their social structure, moving away from living in groups of a few individuals to living in societies. Music may have facilitated this evolutionary shift. For Isabelle Peretz, a researcher at the University of Montreal in Quebec, music does much more than "sculpt our brains."[237] It also triggers emotions and addresses our need to belong. In that way it is an extraordinary "means of communion," increasing our prosocial behaviors and our propensity for altruism. For example, as Peretz points out, "there is no religious ceremony without music and singing." Music was undoubtedly an excellent social glue that united humans around beliefs and contributed, in its own way, to strengthening our societies.

4

Vices and Virtues

Morals are in the head, and morality in the heart.

PIERRE-CLAUDE-VICTOR BOISTE,
LE DICTIONNAIRE UNIVERSEL DE LA LANGUE FRANÇOISE, 1800

I have often been asked if animals have morals, as if Good and Evil were absolutes—independent, fixed, and universal entities. We have created rules for living together in community to which we subscribe so strongly that we believe not only that these rules exist but that they are natural laws that should also govern the rest of the living world.

Yet human morality is a plural concept, generated by our brains and subject to culture. Its diversity in space and time is there to remind us of its relativity.

The Animal Roots of Morality

In Peru, at the beginning of the 16th century, the Chimu people had the custom of jointly sacrificing children and young llamas, using the sacrificial technique called the *ch'illa*, which involved making an incision in the sternum and tearing out the heart of the victim as an offering to the gods. Most likely this sacrifice was to appease the gods during severe weather, especially during the arrival of the climatic phenomenon now known as El Niño.* On the sites of Huanchaquito–Las Llamas and Pampa la Cruz, at least 269 children and 466 llamas were exhumed.[238] The Chimu were not a small group of primitive people with atypical customs but a refined civilization whose architecture and artisanal creations have not ceased to fascinate anthropologists and archaeologists. These sacrifices, which today we find unacceptable, were seen as necessary for the well-being of the community.

Throughout history, cannibalism (another practice that we find abhorrent) has not only been an act of survival during terrible famines but also a long-standing practice of many tribes. When done for food, tribes would make no distinction between animal meat and that of a human belonging to a foreign group. As part of funeral rites, tribes practiced cannibalism to appropriate the soul of a deceased relative. We consider these behaviors to be extremely barbaric because we look at them through the lens of two millennia of monotheism. We lack impartiality because we cannot rid ourselves of the influence of the values we have been taught.

Until 1981 France's customary judicial sentence was to apply the law of an eye for an eye, executing murderers by guillotine. Former minister of justice Robert Badinter recalled that putting someone to death involved "cutting a living man in two"† as punishment for heinous acts. He fought a long battle to abolish this practice, which until the 1980s had been considered moral. Even today the death penalty is still practiced in the United States, where attempts have

---

* El Niño (literally "the child," a reference to the Christ child) is a term used to describe a climate pattern of unusually warm surface waters near the coast of South America. Linked to a cycle of atmospheric pressure variations between the east and west parts of the Pacific and coupled with a cycle of the ocean current along the equator, it is likely to cause very heavy rainfall in Latin America that leads to flooding with disastrous consequences.
† Robert Badinter explained to juries, trial after trial, "You don't realize what you're being asked to do. What is the guillotine? It's cutting a living man in two."

been made to make it more "humane" through administering a series of injections to anesthetize and paralyze before the lethal injection. If there is a shortage of lethal-injection drugs, however, electrocution, gas chamber, and firing squad remain "acceptable" methods. The "rights of man," which seem natural and universal to us now, took root in 1776 in the United States in the unanimous declaration by the thirteen colonies, as Patrice Rolland, associate professor of public law at Paris XII University, points out.[239] The Declaration of the Rights of Man and of the Citizen from 1789 was largely inspired by these texts, but it was only during the first half of the 20th century that human rights became an international topic. Today these principles are so firmly anchored that we think of them as physical laws governing reality, even though they did not exist three centuries ago. In reality, our system of beliefs is intimately—and unconsciously—linked to the social group in which we grew up. We tend to believe that our views were forged by reason, and we lose the ability to clearly see that we inherit most of our views from our culture.

Indeed, morals do not only vary over time but also according to place. In a report illustrating the moral differences between forty countries, the Pew Research Center presented data showing that, depending on where we live, we do not perceive abortion, homosexuality, or the death penalty in the same way. The judgments we make are largely impacted by our beliefs. In many of the forty countries, with the exception of Canada and those in Western Europe, the majority of respondents believed that God is essential to morality,[240] especially in Central Asia and West Africa, and even in the United States, where more than the majority of Americans shared this view. For theists, God is not only the arbiter of justice but also the author of morality.[241] This has a major implication: theists' rules of conduct are governed by their beliefs. Atheists, on the other hand, do not believe that it is necessary to believe in a god to be a moral person. Stripping morality of its universal character, atheists consider it to be subjective and dependent on culture. Still, depending on the country in which they grow up or their education, atheists also build their own value system.

What fascinates me about these observations is that despite their divergence across time and space, the moral standards of theists and atheists intersect in different places. Whether we are believers or not, no matter if we live in cultures that clash with each other, the vast majority of us share moral *intuitions*. For example, we universally abhor unjustified torture or murder and disapprove of injustice, as if we were wired to have these reactions. Which brings us to the following question: given that human morality is a mental construct, did it nevertheless originate from a *biological disposition* that we adapted in the form of shared representations according to our cultures?

The answer to this mystery lies in a combination of prosocial skills, among which is empathy. This capacity is defined as the ability to understand and feel the emotions of others. It is something that continues to captivate the scientific community. This capacity bestows humans with an innate "moral sense," which I distinguish from "morality." With the exception of a neuropsychological anomaly, our motivation to help others, our desire for justice, and our aversion to torture all seem to be inscribed in us. In my opinion, it was when the size of our ancestors' groups increased and they began to pool their knowledge that their prosocial dispositions were put to use and became expressed as shared representations. Morality is, in my view, a concept that took shape in preexisting biological capacities. Produced by the "melting pot of brains," morality created norms to obey and regulated a growing number of interactions between humans. It worked so well that it was transmitted from generation to generation and was adapted according to civilizations and beliefs.

Empathy, which has been widely studied in humans, is among these dispositions. It is split into two components: emotional empathy, which allows us to feel an emotion when we observe someone else's emotional state; and cognitive empathy, which allows us to recognize another person's mental state. The first involves the subcortical area and temporal lobes, while the second involves the prefrontal cortex. Activation of these different zones, in a sort of cerebral symphony, produces a response that our predecessors once called "sympathy." An initial study conducted by the team of American child psychiatrist Carolyn Zahn-Waxler showed that babies aged eight to ten months attempt to comfort distressed individuals in their presence. This consolation behavior continues to increase with age, particularly during the second year of life.[242] Australian researchers from Charles Sturt University corroborated these results. In a group setting of crying babies, babies as young as eight months looked more frequently at the mothers of distressed babies, as if expecting the mothers to console them.[243] From birth, we are inclined to feel empathy. It seems, however, that we are not born equal in our capacity to feel or decode the emotions of others, since our capacities vary according to our genes (with women presenting a higher degree of empathy than men[244]) and the environment in which we grew up. Autistic people, for example, experience difficulties with cognitive empathy in some circumstances but not with affective empathy,[245] the way patients with bipolar disorder do.[246] People with schizophrenia may experience difficulty deciphering emotions from facial expressions,[247] while patients with major depressive disorder—those with a bias toward negative emotions (the tendency to see the glass as half-empty rather than half-full)—may overinterpret certain emotional states since they are particularly sensitive to sad facial expressions rather than happy ones.[248]

Since empathy seems encoded in our genes, could it be that it is not uniquely human? In other words, could the "moral sense" that we thought was specific to humans be more universal? The fifty-thousand-year-old skeleton of Shanidar I, a Neanderthal found at the Flower Burial site in Iraqi Kurdistan, leaves little doubt on the subject. With multiple injuries prior to his death, Shanidar I suffered from visual impairment, severe hearing problems, degenerative joint disease, and had an amputated forearm. According to anthropologists Erik Trinkaus at the University of Washington and Sébastien Villotte at the University of Bordeaux,[249] the poor guy managed to survive for years—at a time when survival was a challenge—because he had been able to count on the help of his peers.

Therefore, empathy is not specific to humans, or at least not to *H. sapiens*. The discovery of Shanidar I suggests that at the very least empathy appeared in the common ancestor of *H. sapiens* and Neanderthals. Could it be then that we can trace its origins back even further in the evolutionary tree? If so, it would give nonhuman animals the biological basis for morality and greatly change the way we think about them. Let's remove the suspense—according to Jean Decety, professor of social neuroscience at the University of Chicago, empathy appeared as soon as the first mammals emerged, two hundred million years ago! In 1872, when the book *The Expression of the Emotions in Man and Animals* was published, Darwin already had this intuition about empathy. He proposed that human beings were not the only ones to perceive and feel the emotions of others, and that this capacity was present in all mammals and probably even in other animal classes.

What does science say on the subject? Some sixty years after Darwin's book was published, an American professor of psychology named Russell Church recorded the reactions of rats trained to pull a lever to obtain food. Despite the food reward, all the rodents stopped manipulating the lever when they noticed that their action caused a negative reaction, through an electric shock, in their fellow rats.[250] These results were subsequently corroborated by numerous other experiments.[251] Rats are eminently social creatures who definitely do not deserve the treatment we reserve for them.

For neuroscientist Cristina Gonzalez-Liencres and her team from the division of cognitive neuropsychiatry at Ruhr University Bochum in Germany, empathy emerged during evolution concurrently with the development of parental care. At this time the attention given to the young began to diversify, extending beyond the simple purpose of feeding, in order to increase the species' chances of successful reproduction.[252] According to this hypothesis, empathy developed as parental care evolved into complex behaviors. The scientists point out that the evolutionary advantage of developing empathy is not limited to direct offspring since giving warmth, comfort, and protection to one's offspring provides another benefit, and a good one at that: by preparing their young for their future parenting role, the chances of survival for the next generation increase. In other words, the more care that is given in raising the young, the more likely they will be to provide the same level of care to their own newborns when they reach adulthood.

This evolutionary claim opens the door to the existence of empathy in other animal classes beyond mammals and more precisely in animals with complex parental behaviors. Given the importance of parental investment in birds, does this animal possess the neural foundations of empathy like mammals do? While we have far less knowledge about these descendants of dinosaurs than we have about mammals, a study carried out by English researchers from the University of Bristol sheds light on what some ornithologists already suspected: mother hens (no pun intended!) show physiological signs of stress when they perceive that their chicks are threatened.[253]

So then, could it be that the capacity to help others is more universal than we thought? In 1899 Russian zoologist Peter Kropotkin saw altruism as a key factor in evolution.[254] According to him, nature and society cannot be considered to be mere gladiator fights from which the strongest emerge victorious. He saw the process of mutual aid as being as important as natural selection. He opposed the reductive vision of Darwinian theory, whereby "man is wolf to man,"* and reminded us of the importance of altruism in the evolution of all social animals. Darwin himself had sensed the importance of empathy in the development of moral qualities, believing that "parental and filial affections . . . lie at the basis of the social affections."[255] This does not contradict natural selection, since it evokes the adaptive advantage that the capacity for altruism represents. Darwin continued, writing that primeval humans "would have felt uneasy when separated from their comrades, for whom they would have felt some degree of love; they would have warned each other of danger, and have given mutual aid in attack or defence." The nonexistent contradiction between altruism and natural selection has also been a theme in sociobiology. The American biologist Edward Osborne Wilson sought, in his book *Sociobiology: The New Synthesis*,[256] to answer the following question: "how can altruism, which by definition reduces personal fitness, possibly evolve by natural selection?"

In 1962 a team of American researchers designed an imaginative experiment to assess the existence of altruism in animals. They suspended a rat and a plastic block separately in the air using harnesses (like Tom Cruise in *Mission Impossible*). An observing rat had the option to lower either its fellow rat or the inanimate object by pressing a lever. All the test subjects operated the lever much more frequently to help the distressed rat than the plastic block.[257]

---

* The expression *Homo homini lupus* ("man is wolf to man") is taken from the book *Leviathan* written by the British philosopher Thomas Hobbes, in which the pre-social individual was described to have been competitive, individualistic, and aggressive.

Capuchin monkeys are also sensitive to the well-being of their companions. In a food sharing task, proposer monkeys could choose to deliver marshmallows (a large reward) or celery (a small reward) to receiver monkeys. The study found that they regularly chose to give the sweet rather than the vegetable reward, even though they did not gain anything for themselves.[258] In elephants, numerous observations have revealed an inclination for altruism, both in situations of danger and when faced with the death of a fellow elephant.[259] Their capacity for empathy was investigated in detail by an English team that collected reports of all their altruistic behaviors over a period of thirty-five years.[260] Forming coalitions, offering protection and comfort, finding baby elephants and caring for them, helping individuals who have difficulty moving around, and removing foreign objects attached to the bodies of others are included on the long list that the researchers managed to compile. What about man's best friend, the dog? Scientists at Goldsmiths College in England drew inspiration for their experiment from an experimental protocol originally created to test empathy in human children. In the presence of a stranger pretending to cry, the test animals, rather than moving closer to their comfort zone (a familiar human), would approach, look at, touch, and lick the unknown human. While the authors discussed the different interpretations of their results (including the phenomenon of emotional contagion*), they also argued in favor of the existence of a genuine capacity for empathy in these wolf descendants.[261]

Beyond their altruistic behaviors, humans also seek to ease the pain or sorrow of others. But are nonhumans capable of this? Different research teams have identified consolation behavior toward peers in bonobos,[262] chimpanzees,[263] and gorillas,[264] further shrink-

---

* Emotional contagion is the transfer of emotions from a sending subject to a receiving subject. It is distinguished from empathy, which implies the existence of an emotion and a cognitive capacity to perceive the reactions of others.

ing the difference between great apes and humans. In a study conducted on elephants in Thailand, researchers Joshua M. Plotnik and Frans de Waal observed increased physical contact (especially trunk to trunk) and vocal communication between elephants following a stressful event.[265] Plotnik described how the type of vocalizations used to reduce the anxiety of a peer sounded to him, saying it was like "'Shshhh, it's OK,' the sort of sounds a human adult might make to reassure a baby." Quite different from their Hitchcockian image, crows also comfort one another. Unlike primates, these highly intelligent birds do not reconcile with their opponents following a conflict, but the initiators as well as the targets of aggression were observed being consoled by social partners not involved in the fight.[266] The wolf, whose monstrous reputation fills children's storybooks, appeared in a new light when ethologist Pierre Jouventin from Montpellier, France, conducted an unusual experiment in the 1970s, adopting one of these wild canids. The many altruistic behaviors of his wolf undermined the myth of the "big bad wolf." Researchers Elisabetta Palagi and Giada Cordoni from the University of Pisa in Italy corroborated these results, showing that wolves that were not directly involved in a conflict displayed affiliative behaviors toward the victim of the aggression—in other words, consolation behaviors.[267] And their descendants are not ones to be outdone. Dogs also respond to the distress of other dogs by exhibiting affiliative behaviors,[268] which corroborates the genuine existence of empathy in these animals.

Could it be that consolation behavior, which was thought to be found only in species with higher cognitive abilities, appears in smaller animals? One team looked at prairie voles, which are known to form monogamous pairs in which the mother and father raise their young together. The scientists separated members of the same family and included individuals who did not know each other. They subjected some of the prairie voles to stressors, either by isolating them completely or by giving them small electric shocks. Once reunited with a familiar partner, the vole not subjected to a stressor increased its grooming of its stressed partner. If the voles were strangers, this grooming would not take place. Are these rodents showing consolation behavior? That is the conclusion of these researchers from the Emory National Primate Research Center in Atlanta.[269]

When a chimpanzee is under stress, one of its group members can offer support. This altruistic ape will most often show no empathy for chimpanzees of another group, however, and will even display aggressiveness. Humans also instinctively tend to act altruistically, but there is a major difference from animals. We can *blindly* empathize with humans we do not know. Blood donation is the most eloquent example of this. We do it without knowing the people who are likely to benefit from it. This difference in magnitude is intimately related to our belief systems, which have encouraged interconnected groups and the use of our skills, including empathy. As the philosopher Georges Chapouthier underscored in his excellent article "L'homme, un pont entre deux mondes: nature et culture" (Humankind, a bridge between two worlds: Nature and culture), human morals are "an analytical and discursive conception" of moral sketches observed in nonhumans.[270] Still, like animals, we are certainly naturally inclined to help another human being, but we find it much more difficult to show empathy toward other species. Of course, our pets—whom we have modeled in our own image—are the exception! And our difficulty in expanding our empathy increases as the species moves further away from us phylogenetically.[271]

While *H. sapiens* excel in feeling and expressing empathy, they also exhibit aggressive behavior that can be terrifying. Psychiatrist and hospitalist Alexandre Baratta explains how a high degree of empathy can lead to physical or verbal aggression in emotionally invested people.[272] But aggression also occurs when there is a lack of empathy. A moderate example would be when we vilify another driver while we are safely in our own vehicle. But in an extreme form, a lack of empathy can lead to horrors, such as murders, genocides, and wars. So, do *H. sapiens* stand out in the kingdom of beasts because of our aggressiveness?

Is Evil Specific to Humans?

Among nonhuman mammals, hostility between rival groups is quite widespread,[273] but it rarely leads to death. The frequent fighting between males is most often limited to intimidation behavior. While certainly frightful, it is rarely fatal. There is one exception, however: our closest cousins, the chimpanzees! Ethological studies have shown animals to be capable of forming complex *political* alliances. English primatologist Jane Goodall made a major discovery on this subject when she revealed an unsuspected *dark* side in chimpanzees.

In 1974, when Goodall was studying the behavior of chimpanzee colonies in Gombe, Tanzania, she observed a social divide between two groups in one of the communities. The first group, called the Kasakela community because they occupied the north part of the park bearing this name, was composed of eight adult males and twelve adult females, as well as their young. The second group, called the Kahama community, consisted of six adult males, an adolescent male, and three adult females. The hostilities began in an extremely violent way when a male from the Kasakela group killed Godi, a male from the Kahama group.[274] The rage of the Kasakelas continued to plague the Kahamas for the next four years, during which time six more males were killed. As for the Kahama females, two disappeared and three were beaten by a gang of violent males. The end of this "four-year war" resulted in the Kasakela community taking over the Kahama's territory. It was a short-lived victory, however, since another community of chimpanzees living nearby managed to scare the Kasakelas away. Goodall recounted her poignant memories of this war in her memoir *Through a Window: My Thirty Years with the Chimpanzees of Gombe*.[275] She recalls, "For several years I struggled to come to terms with this new knowledge. Often when I woke in the night, horrific pictures sprang unbidden to my mind—Satan [one of the apes], cupping his hand below Sniff's chin to drink the blood that welled from a great wound on his face; old Rodolf, usually so benign, standing upright to hurl a four-pound rock at Godi's prostrate body; Jomeo tearing a strip of skin from Dé's thigh; Figan, charging and hitting, again and again, the stricken, quivering body of Goliath, one of his childhood heroes." Jane Goodall is not the only one to be haunted by the bloody images of murders between groups of chimpanzees. American researchers reported similar scenes of violence among chimpanzees in Kibale National Park in Uganda. These primates' fierce battles were instigated by coalitions of adult males, with the sole aim of extending their territory.[276] The areas where the fighting took place corresponded to the land conquered by force.

Are these primates really at "war"? If we define war as being lethal violence organized against members of another group, then the answer is clear. Like humans, chimpanzees have the capacity to wage war.[277] Before the fighting began in Kibale National Park, the males carried out systematic patrols. The location of the corpses confirms the importance of the territory as a motivation to fight: these chimpanzees had breathed their last breath in this coveted neighboring area. These wars were fraught with the terror of infanticide between rival gangs, atrocities also committed by humans. Three such attacks were reported by anthropologists from Ohio University and the University of Michigan in the *International Journal of Primatology*.[278] The researchers recounted how on different occasions, while on patrol, the adolescent and adult males of the Ngogo chimpanzee community attacked the children of a rival gang, killed them, and cannibalized one of them.

Although there are cultural disparities between our ways of waging war and those of chimpanzees, certain similarities are striking. Both humans and chimpanzees ensure that assassinations can be committed by several individuals without major risk to the assailants, and both have motivations for these killings (gaining territory, hierarchical position, access to resources, etc.). In fact, some researchers are now using the "chimpanzee model" to explain the emergence of war in humans.[279]

But aggression in chimpanzees does not only manifest itself when faced with a rival community. American anthropology professor Jill Pruetz and her team at Iowa State University recounted the 2013 murder committed by several males of a member of their own group at Fongoli in Senegal.[280] While the researchers did not witness the massacre as it took place, which was in the darkness of night, they did hear the bloodcurdling cries. In the morning, they discovered with horror the corpse of Foudouko, a seventeen-year-old former alpha male, who had been stripped of his status in 2007 by a gang of young chimpanzees. Condemned to exile and isolation, the pariah regularly attempted to rejoin the group, imposing himself as dominant, which the new alpha males did not like. The research team speculated that if his entrance had been more submissive, the outcome would probably not have been fatal. These lethal attacks recorded in chimpanzees, rare but incredibly cruel, were not linked to a human presence near their communities (as some scientists had presumed) but to a hierarchical tension within the group and probably to intense competition for access to females.[281] But what disturbed scientists the most was how the gang treated Foudouko's body the day after his death. Most likely to make sure they had nothing left to fear, the murderous gang dragged the body across the ground, sniffed it repeatedly, ripped out its genitals, bit it all over, and tore its flesh and . . . ate it!

Murder and cruelty are therefore not unique to *H. sapiens*. And the animal world has not finished surprising us either. Because of their infinitesimal size, we have little empathy for ants, commonly spraying them with insecticides or setting traps that poison entire colonies. The general public learned about these social insects, however, through Maurice Maeterlinck's body of work, for which he won the Nobel Prize for Literature in 1911, and through Bernard Werber's thrilling science fiction novels.[282] Ant colonies, made up of millions of individuals, captivate us with the richness and complexity of their behavior. And there is no shortage of scientific experiments highlighting ants' incredible abilities. Their division of labor, aphid farming, fungus growing, waste management, and traffic control show how ants have elaborate levels of organization. Faced with different problems, these insects are capable of showing great ingenuity. Researchers from Huazhong Agricultural University in Wuhan, China, and a laboratory in Mississippi arranged sugar water in small containers. *Solenopsis richteri* ants adjusted their behavior based on the drowning risks for accessing the sugar water by accumulating sand grains outside and inside the containers to build a structure to function as a syphon,[283] a feat that hardly surprises researchers anymore, given ants' great adaptability.

Beyond ants' collective intelligence, another field of ant research makes the hearts of myrmecologists beat fast: war! Ants, like humans, excel in the *art* of combat. They use many different modes of attack and implement strategic decisions with startling similarities to human military operations. Like our soldiers, ants fight for territory, food, and even to obtain new labor. In fact, some species use their opponents as slaves. By 1810 the Swiss entomologist Pierre Huber had identified species of ants capable of stealing the eggs, larvae, and pupae of rival ants to mature them into obedient workers who would work for them until their death. But while their mission is to care for the pupae of the slave-making ants, some enslaved workers display truly rebellious behavior, killing up to two-thirds of the pupae they are supposed to care for[284]—a revolt carried out well after the defense of their colony had failed!

Those are not the only similarities with human wars. Despite their inclination for altruism within the colony, during conflict, ants show no mercy for the soldiers of the opposing party, dismembering them or tearing off their head or thorax with blows from their mandible. Just like terrorists who believe they are serving their cause by blowing themselves up, *Colobopsis explodens* ants deliberately rupture their gaster in order to release a sticky liquid irritant to kill or repel their rivals.[285] While these kamikaze insects cause surprise with the immediate sacrifice of their lives for the colony, other ants have developed attack techniques of an equally impressive efficiency. Without a stinger, red ants attack their opponents by biting them and injecting formic acid directly into the wound. It is deadly enough to strike down the enemy on the spot!

Entomologists have documented the war tactics of ants. Certain ants claim victory by adopting an all-out offensive strategy, while others use a maneuver akin to siege warfare, monopolizing resources and starving their competitors, who send out scouts to collect information on the location of their enemies. Once they return to the nest, the fight is on! The soldiers go to war. They set off to attack and raid termite mounds for food. A high-risk adventure, the ants tirelessly fight soldier termites with sharp mandibles, and they rarely return home unscathed. Back at the anthill, the injured attackers are taken care of by a remarkably well-organized "health service."[286] First, the caregiving ants triage the wounded soldiers by the degree of severity of their injuries. Leaving their most severely injured compatriots to their fate, the health-service ants give special care to lightly injured ants. During the first hour of their arrival at the nest, the lightly injured ants are examined much more thoroughly than the uninjured survivors. Using their glands, the "nurses" secrete antimicrobial substances that they spread over the cuticle of the injured ants to inhibit the risk of infection and give them an antiparasitic treatment. Without this vital care, the mortality rate of injured ants would increase significantly.

Wars between ants do not always end in a "bloodbath,"* however. Small colonies will often prefer flight over fight. Since they are able to quantify the number of individuals in a foreign colony, ants can gauge the risk of going to war. This type of decision is made according to local rules, a kind of swarm intelligence that is radically different from human military leadership. Moreover, although it is tempting to compare humans and ants, the two species are so far apart from a phylogenetic standpoint that it seems difficult to deduce human behavior in relation to ant behavior.

In any case, if *H. sapiens* are not the only ones waging war, does that mean that warfare has been part of our species' nature since the dawn of humanity? We can understand the ideological stakes of the question—if aggressiveness is genetic and not cultural, it is more difficult to keep in check. Long before our ancestors used gunpowder, they fought with swords and blades. Before these weapons, they fought their battles armed with spears and projectiles. But did this aggressive tendency between groups come about with the appearance of civilizations, or has it characterized humanity since the beginning of time? The origin of collective violence is a thorny question, which archaeological data has shed light on. All the excavations carried out agree on one point: as soon as people settled down, raising livestock and farming, wars sprang up left and right.

---

* To be exact, it would be called a "hemolymph bath" (the equivalent of blood in insects).

The possession of land and resources by certain groups of individuals aroused envy and jealousy. Therefore, until recently, it was difficult to imagine the existence of wars prior to the development of sedentary societies. Two theories clashed: on the one hand were the partisans of Rousseau, who defended the idea that uncivilized humans were not passionate by nature but had been transformed radically by nascent society; and on the other hand were those who believed humans were indeed violent by nature.[287]

A discovery made at the archaeological site of Nataruk (the "place of the vultures"), west of Lake Turkana in Kenya, helps us to take a stance by demonstrating the existence of wars among hunter-gatherers (before they settled into sedentary societies) nearly twelve thousand years ago.[288] The remains discovered in this marshy area had been exceptionally preserved. The analysis carried out on the twenty-seven skeletons revealed a chilling scene. A pregnant woman was found in a seated position with her hands placed between her legs and bound by her assailants. A little farther away, another woman was in the same position as her friend, and her thorax, both knees, and her left foot had been fractured by blunt-force blows. Testifying to the rage of their torturers and their thirst for domination, this seated position was observed in many of the victims. The aggressors had had a field day. At the same site, another woman's vertebrae had been broken by a blunt instrument and her skull fractured. One of the men of the tribe, whose knees had been fractured, bore the marks of blows to the head. These twenty-seven people died of their injuries right where they lay. Their abandoned bodies remained forever in the position in which they had been slaughtered. The nearby lagoon covered them, the carnage falling into oblivion, until twelve thousand years later when erosion revealed the horror. Obsidian (a volcanic rock not present in the Lake Turkana region) was found embedded in some of the bones, evidence that the assailants were nomads from a distant land.

Another chilling vestige of our warlike past was discovered in Sudan. Fifty-eight skeletons of men, women, and children were found at the Jebel Sahaba cemetery, dating back approximately fifteen thousand years.[289] Investigations at the scene of the crime revealed that nearly half the victims had died following an extremely violent attack, with bone evidence showing they had been pummeled in the head, pierced by spear points, or had bones broken by stone projectiles, some of which were still embedded in the bones. Unlike the Nataruk massacre, where the bodies had been abandoned, the murdered bodies of Jebel Sahaba had been buried. This suggests that the attackers were a group of "pre-settled" hunter-gatherers who practiced funeral rites, because, in this geographical area, no trace of a village has been found dating to this period of history.[290] The murderous tribe had probably settled on the right bank of the Nile to enjoy the abundance of resources during the summer, fall, and early winter. In the depths of winter, when stomachs began to cry out for food, competition became intense. This would lead to deadly raids on rival tribes—killing to avoid being plundered of one's meager resources.

As we can see, violence preceded the birth of civilizations. American anthropologist Brian Ferguson of Rutgers University in Newark, New Jersey, points out, however, in his paper "Violence and War in Prehistory," published in *Troubled Times: Violence and Warfare in the Past*, that while wars have punctuated our history, not all human groups and societies have always shown violence.[291] Many tribes traversed the centuries without engaging in intense fighting nor brutalizing their rivals. And aggressive behavior does not always lead to war. In *On Aggression*,[292] Austrian biologist Konrad Lorenz, one of the founding fathers of ethology, explains that, for the most part, evolution in animals has produced ingenious mechanisms to direct aggression toward harmless paths. He notes that *H. sapiens'* development of weapons disrupted this, however, because "the man who presses the releasing button is so completely screened against seeing, hearing or otherwise emotionally realizing the consequences of his action."

Before *H. sapiens* appeared on the scene, were their most distant ancestors also violent? In 1953 the famous Australian anthropologist Raymond Dart, discoverer of the Taung Child (the first fossil ever found of *Australopithecus*), put forward the hypothesis, based on discovered bones and weapons, that *Australopithecus africanus* did not only hunt other mammals for food but also killed individuals of their own species.[293] According to Dart, human wars evolved from aggressive instincts already present in *Australopithecus*. He described them as being carnivorous creatures that "seized their prey alive, beat them to death, dismembered and skinned the bodies, quenched their fierce thirst with the still hot blood of their victims and voraciously devoured the quivering flesh." *H. sapiens*, their worthy heirs, then spread from Africa and across Eurasia, eliminating all other species of bipedal apes on their way. Despite a number of critics, the "killer ape" hypothesis exerted considerable influence. Ultimately, the theory was abandoned in the 1980s when scientists found that the *Australopithecus* bones and the animal horns supposedly used to make weapons exhumed nearby were not evidence of a massacre between *Australopithecus* groups but instead were the remains left by predators, such as leopards, that had devoured both *Australopithecus* and the horned animals.[294] We tend to forget this, but at that time, "prehumans" were more often prey than hunters!

Many years later in 1999, the English primatologist and anthropologist Richard Wrangham from the department of anthropology at Harvard University formulated an alternative hypothesis, different from that of the "killer ape" but that also claimed war to be rooted in animal origins.[295] While some researchers have long thought of war as specific to humans, given that they created weapons and a patriarchal ideology, Wrangham pointed out that the existence of coalitions between adults, for the purpose of killing members of a neighboring tribe, is not particular to humans because it also exists in chimpanzees. According to him, war originated in "tendencies" that already existed in prehumans. A "territorial instinct" appeared in the common ancestor of chimpanzees and humans that explains the emergence of intergroup fighting in these two species. The patrolling behavior of young male chimpanzees that leads to killings in order to take control of new resources provides evidence of this. Even though it is enlightening to analyze behavior from the viewpoint of the ancestral past, for the English archaeologist Nick Thorpe of the University of Winchester in England, attributing human wars to our "territorial nature" would be a bit hasty.[296] The Dutch primatologist Frans de Waal adds that it is a debatable choice to use the chimpanzee as an explanatory model rather than another one of our close cousins, such as the bonobo,[297] especially since bonobos, unlike chimpanzees, do not practice infanticide, cannibalism, or have intergroup conflicts. Instead, they prefer to resolve tensions through sex!

Since then, several genetic research studies have been carried out on these two species. What do they say? In 2005 complete genome sequencing of chimpanzees revealed DNA with nearly 99% similarity to humans',[298] suggesting that they are our closest cousins. But another study in 2012 by a team from the Max Planck Institute ultimately showed that, from a genetic point of view, bonobos are as close to us as chimpanzees.[299] Additionally, another study pointed out that a number of genes shared between chimpanzees and humans can be expressed differently.[300] Therefore, it becomes difficult to infer the social behavior of *H. sapiens* from a common ancestor of chimpanzees, bonobos, and humans, as the few genetic differences and variations in the expression of the genes we have in common likely account for major behavioral differences, particularly aggressive behavior.

Considering what we currently know, we must admit that we are still at the conjecture stage concerning the emergence of war among our four-legged ancestors. But one thing is certain: from the beginning of *H. sapiens*' time on Earth, they have exhibited violent behavior. This assessment might have condemned us to fighting eternal wars, if not for the fact that contrary to our aggression, we have developed high degrees of empathy and intelligence, which allow us to overcome and control our darkest instincts. The idea of a nonkilling society is not a naive myth. American anthropologists Robert Knox Dentan[301] at the State University of New York at Buffalo and Clayton and Carole Robarchek[302] at Wichita State University in Kansas studied the customs of a nonviolent people in Malaysia called the Semai. Among this aboriginal people, nonviolence is instilled from childhood and plays a major preventive role in inhibiting aggressive behavior in adulthood. The Semai teach all members of their group to respond to situations that may trigger violent behavior, such as frustrating stimuli, with a fear response that inhibits criminal acts.[303] For the individuals of this society, violence seems frightening and meaningless. When tensions arise, they are eased through various means, including discussions mediated by elders. The Semai's strong cooperative ties and extremely low crime rate are also a reflection of their worldview. Anything external to the community is seen as dangerous and potentially hostile. Their only refuge is therefore the community in which they live—a protective bubble that must be maintained by preserving peaceful internal relations. In this context, violence and nonviolence no longer appear as inevitabilities but as two behavioral potentialities. The Semai show us that far from being an unavoidable evil, violence in our societies can be influenced by our educational and social model. The question is no longer about whether it is in our nature to make war, but about which worldview inspires us and what sociopolitical implications that has.

While today *H. sapiens* are entirely aware of the disastrous consequences of violence (which we still have some difficulty curbing), we were not aware until a few decades ago of another evil that has been within us for thousands of years: our tendency to destroy living things.

An Ecocidal Primate

The unparalleled adaptability of *H. sapiens* certainly allowed them to step into new biotopes, but their tread was not so light. Each time they came to a new place, the local fauna fell to ruins. Experts once explained this drastic reduction in the number of animal species by climate-change phenomena, but the dating of various climate events reveals a strong concomitance between the arrival of *H. sapiens* in a given geographical area and the disappearance of megafauna.[304] While it is simplistic to say that hunting instantly led to the extinction of prey, history reveals to us that our arrival in different parts of the globe did precipitate the eradication of hundreds of species through the combination of direct (hunting) and indirect (competition and habitat alteration) pressures. *H. sapiens* reached the top of the food chain by developing social cooperation and cognitive faculties. Unlike other top predators, however, whenever food ran out, they began to conquer new territories and hunt new prey in new ways, especially by allying with the wolf, which was devastating for other species.

Let's get back aboard our DeLorean and begin a new journey in time, this time to thirty thousand years ago on the American continent. To the north is a herd of immense pachyderms, enthralling with their huge size and 10-ton weight. Here we are in front of the majestic Columbian mammoths, whose curved tusks make formidable weapons. If we cross a small stretch of the Pacific Ocean to reach the Channel Islands of California, a completely different spectacle awaits us, namely a "hobbit" version of the Columbian mammoth, its shrunken size due to island dwarfism. Now let's return to the mainland where a rodent as big as a bear, measuring 3 meters (nearly 10 feet) long and 2.5 meters (over 8 feet) high, and endowed with pointed incisors and a rounded tail is enjoying something to eat. This giant beaver does not build dams but relies on swamps to survive. Let's travel a bit south to South America where we come face-to-face with a beast more than 3 meters (10 feet) in height, whose impressive claws and strong jaws make us quiver in fear, not to mention its long tail that allows it to stand up easily and makes it look like a giant. This monster is actually a calm herbivore called the giant sloth who feeds on various foliage. Not far away, we spot the thick shell of a creature weighing almost 2 tons, but it hardly scares us as it is quite slow-moving. This is the giant armadillo. Relieved, we continue our stroll in this prehistoric world. Suddenly, not far behind us, we hear fearsome panting. With just enough time to climb back into our DeLorean, we steer clear of *Smilodon*, the saber-toothed cat that has been present on Earth for two and a half million years. All these mammals, which survived three extremely difficult glacial periods and endured for centuries, were annihilated with the arrival of *H. sapiens*. The few ground sloths that managed to survive on the Caribbean islands until 5 thousand years BCE were wiped out as soon as our ancestors found a way to cross the sea.

Over in Australia in the unusual world of prehistoric marsupials, we see the same scenario play out. Similar in size to a hippopotamus, the giant wombat, which made its entry into the world 1.6 million years ago and carried its young in its ventral pouch like koalas do, was doomed to oblivion when *H. sapiens* colonized its island home. Giant species of kangaroos, wallabies, and koalas met this same deadly fate. The descendants of the dinosaurs were not spared either. Farewell to the *Dromornis*, which were among the largest birds of all time, and the *Bullockornis*, which measured up to 2.5 meters (about 8 feet). Here again, climate change did nothing to help the fate of these animals, but the pressure exerted by aboriginal hunters and the change in landscape these people brought about profoundly upset the ecosystem, precipitating the decline of these giant species.

And history repeated itself, over and over again. In each colonized habitat, *H. sapiens*' arrival led to the extinction of hundreds of species that had managed to thrive for millions of years. Each plot of land explored by our ancestors was relentlessly emptied of its biodiversity. The dodo, endemic to Mauritius, is the most emblematic recent example. This bird with a distinctive physical appearance lived in the forests and plains of Mauritius and Réunion. About 1 meter (just over 3 feet) tall, it sported blue-gray plumage, small atrophied wings that were yellow and white, and a large hooked beak whose end was decorated with a blue spot. Its head had two prominent folds at the base of the beak. Rather wacky looking, the dodo was discovered when Admiral Wybrand van Warwijck visited Mauritius in 1598. Even though it was well known that the dodo's meat had an unpleasant taste, the dodo was easy prey. The poor flightless bird had not developed any fear of humans. It was hunted to provision ships' crews. But when people began living on the island, the destruction of forests and the importation of nonnative animals, such as dogs, cats, rats, pigs, and crab-eating macaques, decimated the dodo population to the point that they were permanently wiped off the map. In the 19th century, new archaeological discoveries rekindled interest in these birds, which became iconic thanks to the pen of Lewis Carroll and his work *Alice's*

Adventures in Wonderland. Symbols of extinction due to human presence, the stuffed dodo models in natural history museums around the world continue to arouse emotion.

Can we blame our predecessors? *H. sapiens*, like any animal, were driven by a will to live that led them to kill for food, to colonize new territories, and to fight for their survival. There is nothing diabolical to be found in these biological designs. But, unlike other animals and even other human species they coexisted with for millennia, *H. sapiens* managed to join their forces through the "melting pot of brains." Biodiversity still had a chance to recover in places where it was up against small groups of humans, but nothing could withstand the steamroller of human societies. But how can we fault our ancestors for ransacking the planet when they were unaware of the ecological issues they were causing? It is only in the last two or three generations that, thanks to science, we have been able to realize the significance of the situation, especially as the ecological issues have exponentially increased since the industrial revolution. We are now confronted with the greatest challenge of all time: to reduce our disastrous impact on the earth while our exponentially growing population is drastically accelerating the decline of biodiversity. Faced with this alarming predicament, extending our empathy to the rest of the living world seems to me a necessity. This moral shift is already taking place, but it can only be fully achieved by radically reforming legislation, particularly around the environmental rights of wild species. Legislation must be based on science and not motivated by human desires. Only then will we be ready to fight humanity's greatest battle, the battle to save our planet.

5

Sex Machine?

Love at first sight is easy to understand; it's when two people have been

looking at each other for a lifetime that it becomes a miracle.

SAM LEVENSON
(AMERICAN HUMORIST)

Escaping all reason, love at first sight shatters our world as we know it and makes it orbit around a new center, our romantic partner. There we are with sweaty palms, butterflies in our stomach, the temperature rising a notch. If Cupid's arrow can put us in this state, it is because feelings of love are rooted in the oldest parts of our brains, the parts over which we have no control. We can understand love at first sight from a biological standpoint. Functional neuroimaging has revealed its headquarters in the basal ganglia of the cerebral cortex. Love at first sight produces an influx of blood to twelve different areas of our brain[305] that are specifically involved in cognition and emotions. Next these areas release a cocktail of molecules related to attachment, the reward circuit, and alertness. Uncontrollable biological fireworks! Visual, acoustic, and olfactory signals instantly reach the brain, triggering a series of emotional reactions, which explains this feeling of intoxication mixed with sensual ecstasy.

Humans were long considered to be a rather "visual" species, having lost so much of our sense of smell over the course of evolution that we were considered a "microsmatic" species.* That is, until a discovery disproved our knowledge, as is often the case in science! In an article entitled "Poor human olfaction is a 19th-century myth," the American researcher John McGann revealed that the human olfactory bulb, despite being small, allows us to detect and discriminate many more odors than we thought.[306] In 1995 the Swiss biologist Claus Wedekind at the University of Bern and his team highlighted the importance of olfaction in choosing a romantic partner. Interested in a set of genes involved in immune-system response and reproduction called "major histocompatibility complex" (MHC), they hypothesized that MHC influences the composition of our body odors. To find out for sure, the scientists asked male university students to wear the same T-shirt for two consecutive nights. The following day, the female students were asked to rate the scents of six different T-shirts. All preferred the scents of the men who differed the most from them genetically, with dissimilar MHC. All, that is, with the exception of women taking birth control pills, whose preferences were reversed.[307] To explain this exception, the researchers postulated that under the influence of steroids normally released during pregnancy, women must be drawn to the odors of genetically close individuals because they are more likely to help them. In the long term, taking an oral contraceptive could have a negative impact on the stability of a couple, since the woman would be less attracted to her spouse's odor once no longer taking birth control. Scientists also warn about the possible role of perfumes and deodorants in disrupting the detection of body odors relevant to reproduction. Psychologist Jessica M. Gaby at the Monell Chemical Senses Center in Philadelphia tempered this

* "Microsmatic" refers to a vertebrate whose olfactory bulb is poorly developed or almost absent.

last hypothesis, however, by showing that humans are perfectly capable of detecting body odor even when mixed with perfume.[308]

One thing is certain; *natural* odor plays a key role in choosing a romantic partner. It encourages us to opt for an individual who is sufficiently distant from ourselves genetically, at least when we are not under the influence of synthetic hormones like those found in contraception. This biological phenomenon is transcribed in humans by a cultural norm—incest taboo, which is found in the majority of societies. In fact, the tendency to choose a genetically distant romantic partner is found in all mammals (such as primates[309] and rodents[310]), fish,[311] reptiles,[312] and even birds,[313] providing proof that when love strikes, we are subject to the same mechanisms as the entire animal kingdom.

The Trappings of Seduction

But why does natural selection favor choosing partners who present a strong genetic dissimilarity? This is an extraordinary evolutionary strategy to limit the risks of reduced biological fitness and to increase the chances of survival in offspring; or in other words, to ensure inbreeding avoidance and genetic variability. As for MHC-related odors, these are perceived unconsciously. In fact, the olfactory receptors send projections to the neocortex by both conscious and unconscious processes and then to the limbic system, which is involved in the processing of emotions. This explains why a smell can be associated with both memory and emotion. Animals, including humans, avoid inbreeding by mating only with unrelated individuals that are sufficiently different from them genetically to give birth to offspring that will have a higher chance of survival than the offspring of a related male and female. In fact, most vertebrates are able to distinguish between individuals, mainly on the basis of body odor, that have a genetic link with them and those that are not related to them. To understand this phenomenon, Marylène Boulet and her collaborators at Bishop's University in Sherbrooke, Quebec, looked at the possible relationship between genetic similarity and odor secretions in a captive population of lemurs.[314] Using a gas chromatography technique, the researchers analyzed the volatile compounds secreted by the animals' genital glands. The results were revealing. The "odor profiles" of related individuals were more similar than those who were not related. Even more remarkably, despite the differences between the male scrotal glands and female labial glands, this odor-profile similarity between individuals that were genetically related existed even when they were not the same sex. There are additional mechanisms for inbreeding avoidance too. In mammals in particular, physical distancing observed during weaning allows the young to *take flight* and move to new territories, which greatly reduces the risks of later mating with their parents or siblings, and increases the chances of meeting new partners who are more genetically distant.

Other types of odors are involved in romantic choices. The existence of sex pheromones (which are olfactory signals) has been demonstrated in a number of species. Since the discovery in 1959 of bombykol—the sex pheromone released by the female *Bombyx* moth (whose larva is the silkworm) to attract males—by the German biologist Adolf Butenandt, research on insect pheromones has multiplied. And what has been gathered is impressive. To date, more than two hundred sex pheromones for attracting mates have been isolated in lepidopterans (butterflies and moths), and around twenty in dipterans (flies, mosquitoes, etc.) and in coleopterans (including ladybugs and beetles). Not to be outdone are fish and reptiles. Male guppies are more attracted to females that produce a sex pheromone.[315] In reptiles,[316] and in garter snakes (*Thamnophis*)[317] in particular, organic compounds found on the skin of females acting as attractiveness pheromones for males were identified in 1990 by the team of Robert T. Mason, a professor of integrative biology. As far as birds are concerned, a study suggested that, even though no sex pheromones have been identified yet, quail modify their sexual behavior when deprived of olfactory inputs.[318] In mammals, few experiments have enabled these compounds to be formally characterized. The pheromone ESP1 produced by male mice, however, has been shown to increase the female response to their advances.[319]

And humans? Are they also responsive to pheromones? Androstenone, for example, which in pigs is used as a sexual signal, is a molecule resulting from the degradation of testosterone found in human sweat. Andreas Keller at the Rockefeller University and Hanyi Zhuang at Duke University Medical Center observed that not all people perceive this odor in the same way. Some find it offensive as it reminds them of perspiration or urine; others conversely find it pleasant and associate it with a floral smell; and lastly for some it is odorless, and they do not perceive a scent at all. Individuals' sensitivity to this odor varies due to a genetic variation. Keller and Zhuang were able to determine that there are several versions of the gene involved in the detection of androstenone.[320] Other research teams have examined the effect of a similar pheromone—androstenol—on sexual attraction, but the methods used have varied greatly from one study to another and do not have consistent results.[321] In fact, the vomeronasal organ used mainly by mammals to detect pheromones* is vestigial and nonfunctional in humans, which suggests that these compounds might have lost the prominent role they once had in our olfactory communication. This would mean that to choose a romantic partner, the *beast* in

---

* For a long time it was believed that pheromone processing happened exclusively in the vomeronasal organ. The discovery of mammary pheromones in rabbits by Benoist Schaal and Gérard Coureaud's team, however, led to the conclusion that these pheromones were processed by the main olfactory system. Therefore, the respective roles of the two structures are not as clear-cut as they appear.

us no longer expresses itself through pheromones but continues to manifest itself through subconsciously perceived stimuli.

Like humans, animals choose their mate based on a combination of visual, olfactory, and acoustic signals. Let's look at the visual signals first. In birds, for example, evolution has endowed males with remarkable attributes, but these can sometimes be handicaps. This is the case for the peacock. Unlike the female, the male sports a multicolored tail, which makes it visible to predators and reduces its chances of survival. But the beauty and majesty of its *fan* are an undeniable asset during a courtship ritual. With meticulous choreography, for the sole pleasure of its love interest, the peacock spreads its iridescent feathers and vibrates them so that the eyespots (the dark blue spots resembling large eyes) appear as static points that sparkle on a moving background. And the more iridescent the eyespots are, the more inclined the females are to be wooed by their suitors' charms.[322] Charles Darwin had a hard time explaining how evolution could prioritize aesthetics over individual survival. In 1860 he wrote in a letter to a colleague, "The sight of a feather in a peacock's tail, whenever I gaze at it, makes me sick!"[323] But this great scientist who forever revolutionized biology proposed a second mechanism called "sexual selection," which is complementary to the "struggle for existence." Sexual selection favors the transmission of genes to offspring over direct survival. It helps us understand what is called "sexual dimorphism" in biology—the condition in which there are morphological differences of varying degrees between male and female individuals of the same species. Impressive and abundant plumage in male birds allows females to identify strong, healthy candidates. This sexual selection happens, however, not because of the feathers' colors but rather because of their ultraviolet reflections,[324] which these animals can perceive thanks to their ultraviolet vision.

The differences between males and females are pronounced, as evolution has equipped males with multiple aesthetic *jewels* to win over hearts. In mammals, depending on the species, dimorphism is especially apparent in terms of size (with males being larger and stronger than females), coat color, and size of canine teeth. For instance, the lion is larger than the lioness and is distinguished by its thick mane. The stag is crowned with bulky antlers, while the smaller doe has none. Similarly, on average, men are taller, heavier, and hairier than women. Females also have unique traits. Women have breasts, which somewhat impair movement and only serve the function of breastfeeding. In fact, the females of our very close cousins (chimpanzees, bonobos, and orangutans) do not have prominent breasts, except during the breastfeeding period when they become engorged with milk.

So why are women so well-endowed? In his book *The Mating Mind*,[325] the American evolutionary psychologist Geoffrey Miller advanced the idea that the voluminous aspect of breasts has been preserved by a mechanism of sexual selection. In other words, breasts acquired their permanent eye-catching appearance to attract men. While they only have a nursing function in other primates, breasts have become highly erogenous in humans, which corroborates their function as a sexual attribute. But opinions on the matter differ. The Malaysian psychologist Viren Swami and the English neuroscientist Martin Tovée believe that prominent breasts signal an ability to store fat and therefore indicate accessibility to food. In a study of nearly a thousand men in Malaysia and England, individuals from a low socioeconomic background found large breasts more attractive than men from a privileged background. This means that men struggling for survival would be more attracted to large breasts, demonstrating that hunger impacts sexual preferences![326] Given the difficulties the authors had in separating cultural factors from the economic context and the biases that may have occurred in their study, they themselves are uncertain if their conclusions are reliable, however. What do ethologists think about this? In his book *The Naked Ape*,[327] the English zoologist Desmond Morris contends that humans' upright posture on two legs encouraged the adoption of new sexual positions, in addition to those usually observed in animals (such as mounting from behind). According to this claim, evolving to a frontal position would have exerted an evolutionary selection pressure on women, meaning that the roundness of the breasts would have been reminiscent of the roundness of the buttocks, making them a new sexual asset. This hypothesis, however, clashes with observations of our cousins the bonobos who, like humans, use the missionary position and whose females do not have prominent breasts. From the neotenic angle, I can think of one more explanation. Rather than recalling the roundness of the buttocks, it is possible that breast size was selected by evolution because it recalls the moments of intense pleasure experienced during breastfeeding in infancy.

Humans have invented other techniques to seduce partners too. Observing nightclub behavior shows us how people dance with abandon, their creativity knowing no bounds. For that matter, dancing sometimes imitates sexual positions or practices. The lambada in particular is a sensual dance in which the partners' bodies are touching, and the leader places a knee between the follower's legs. A more recent example in popular culture is twerking, during which people evocatively thrust their hips and shake their buttocks.

Those are the *natural* visual signals. But humans also know how to artificially adorn themselves to win over a partner. When *H. sapiens* became naked apes after losing their fur coats—which not only had a role in thermoregulation but also provided information about emotional state (raised hair, for example), health, and social group membership—they started painting their faces and bodies, probably since prehistoric times, to display their social belonging to a group and in the context of shamanic or funerary rituals. In ancient Egypt, makeup was used by both men and women, and had both a spiritual and an aesthetic role. Black eyeshadow, called *mesdemet* in Ancient Egyptian (translated into Arabic with the term *kohl*), not only had disinfectant properties but also was used to accentuate the eyes by making "the eyes speak." Sumptuous cosmetic palettes have been found from the 4th millennium BCE. These were exported to ancient Greece and Rome, where Nero himself painted his eyes with kohl. In the Western world during the 13th century, nobles covered themselves with foundation. Then in the 17th century, aristocrats of both sexes covered their faces with white lead powder called "ceruse" and highlighted the veins on their temples with cobalt blue, so they could prove they were members of the "blue-blooded" nobility and thus make themselves appear more attractive.[328] As Anne-Marie Granet-Abisset, a professor of contemporary history at the Rhône-Alpes Historical Research Laboratory in France, tells us, "Before the French Revolution, men and women wore makeup to represent their social status. Then the bourgeoisie took over and rejected certain artifices that had to do with the aristocracy, such as male makeup."[329] Therefore, it was only after the French Revolution that this practice became particular to women, whereas prior to then cosmetics had been used by both sexes for millennia. Suddenly everything changed, a man's masculinity being called into question if he wore makeup. Beyond the seductive advantage that makeup offers to women who choose to wear it, Granet-Abisset reminds us that wearing makeup also contributes to a "self-image that corresponds to social and cultural codes and norms." Put another way, today not only is makeup used for making oneself appealing to others, but most importantly, it plays a cultural role in beautifying oneself according to the aesthetic standards of a particular society. Depending on the era and cultural customs, a whole armada of ornaments, jewelry, and dress codes has played a role in sexual attractiveness.

Let's move on to the acoustic signals involved in romantic choice. The human voice is another asset in seduction. Nasal or shrill voices tend to be associated with low sex appeal. Among heterosexuals, a masculine voice remains more attractive to women than to men, and the opposite is also true. Heterosexual men are naturally more attracted to feminine voices.[330] A beard and a square jaw in men are also selection criteria for heterosexual women. In fact, the female preference for masculine traits is stronger during ovulation (the phase when women are most fertile) than during the other phases of the menstrual cycle.[331]

But let's get back to love at first sight. Its instinctive quality in humans begs the question of whether this mechanism, which favors the union of two beings, could be more common than we might believe. As Darwin pointed out in his book *The Expression of the Emotions in Man and Animals*,[332] emotions have not changed much over time—at least not for mammals. From a biological standpoint, mechanisms allowing two individuals to be paired for reproduction have a high adaptive value. I would be so bold to promote the hypothesis that animals can also be overcome by their emotions at the mere sight or odor of a congener and, like humans, be overwhelmed by a flood of emotions similar to what we feel during love at first sight.

Love with a Capital *L* Is Not Unique to Humans!

Unlike love at first sight and its ephemeral nature, true love is a powerful emotion that never fades. But how does the magic of love keep two beings united in the long term? The secret lies within the hypothalamus, a region of the brain that is the size of an almond. Located at the base of the brain, it acts as the orchestra conductor by regulating our hunger, thirst, body temperature, and sleep. Beyond these functions, the hypothalamus also secretes compounds with astonishing properties. First of all is oxytocin, whose role is essential in maintaining attachment bonds. Involved in parent-child bonding and released during childbirth, lactation, sexual activity, and orgasm, oxytocin also sets the tone in romantic relationships. It is the same hormone that is found in female prairie voles, which are also monogamous animals.[333] The melody of love seems to play with universal molecular notes. Once they choose a sexual partner, these rodents refuse any other partner thanks to a regular release of oxytocin. During my stays in Guyana, I remember my surprise the first time I observed macaw couples flying two by two, inseparable even in the sky. We also find this unbreakable bond in a small monogamous mouse, named *Mus spicilegus* (which was one of my research subjects), in which the male and female form a close-knit couple, sharing caring responsibilities for their young. This is the case as well with swans, dik-diks (small antelopes), gibbons, wolves, marmosets, and tamarins. Monogamy is certainly not the most widespread reproductive strategy in the animal kingdom, but it appears in a large variety of species. Observations of these loving duos who painstakingly groom each other, help each other, and share the tasks of raising their young leave no room for doubt: a strong bond unites monogamous animal couples.

For a human, when death or separation puts an end to a romance, that individual's whole world crumbles as they embark upon a painful period of "heartache," which includes unbearable emotional distress, loss of motivation, depression, morbid thoughts, and sometimes even suicide. Is it possible that animals also have these painful feelings? We tend to believe that small beings or those living in the aquatic world have little to no cognitive and emotional capacity. Our projections are limited to species that are phylogenetically close to us. An experiment carried out in 2020 by the team of François-Xavier Dechaume-Moncharmont, lecturer in evolutionary ecology at the University of Burgundy, has overturned these false beliefs, however. A small monogamous fish called the convict cichlid can also have a *broken heart*. In this species, the male and female build a nest together, then raise their young together. In their experiment, the biologists used the judgment bias test, which measures emotional valence, to reveal if animals, like humans, can see the glass as half-empty or half-full. In simplified terms, judgment bias shows that when a person is asked to judge a stimulus called an "ambiguous signal," they will tend to rate it negatively if they are depressed or sad and positively if they are happy and optimistic. Dechaume-Moncharmont's team first trained the fish to open boxes by lifting the lids with their mouths. Then they offered the fish two boxes with differently colored lids—one containing a tempting food reward (a type of worm) and the other empty. These animals, which lived as couples, quickly learned that a box with a black lid was of more interest than one with a white lid, which contained nothing. So they all rushed to the black one and opened it very quickly, feasting on the worm inside, while they lost interest in the white one. Then the ethologists repeated the exercise with the ambiguous signal. They offered a box with a gray lid (an intermediary color) to two groups: first, females living with their preferred partner, and second, females that had been separated from their companion and were living with a male replacement. Females separated from their preferred male were not interested in the gray box, exhibiting pessimistic bias, which indicated their negative affective state. On the other hand, those who remained with their lifelong partner did not show any significant response to the gray box. As we can see, *H. sapiens* are not the only ones who experience heartache. These small fish, native to Central America, also feel profound distress at the loss of their romantic partner.[334]

While the proof of a strong bond existing in monogamous animals is now well established, some might argue that humans and other animals differ in their displays of affection. For example, animals do not seem to have the equivalent of our affectionate kisses. Birds, however, do regularly engage in these practices, and not just between a mother and her young for the purposes of feeding but also between adults. As a matter of fact, this behavior can persist in couples, even when the sharing of food is not taking place. Occasionally the male will feed the female in the nest, but he will also do this when no little ones are present and when it is not for nourishment purposes. This beakful of food, of no value to the individual's survival, plays a social role and, more specifically, a role of cohesion in the couple. There are not many studies on the subject, but for those who know how to observe these animals, there is no doubt that this is indeed a way they show their affection. The French naturalist Pierre Rigaux noticed that in the common puffin, also called the Atlantic puffin, the two monogamous partners regularly rub their beaks together, an act similar to our affectionate kisses. In the seahorse, the close-knit male and female couple exchanges elegant bows each morning before going their separate ways for the rest of the day. As for primates, the Dutch ethologist Frans de Waal says that apes "greet each other after a separation by placing their lips gently on each other's mouth or shoulder,"[335] which is more like our platonic kisses. As for bonobos, they do not hesitate to kiss with their tongues!

The kiss is therefore not as unique as it seems. Some scientists have put forward the idea that, like birds, our French kiss is derived from the behavior of feeding our young. Before the arrival of baby food in jars, humans would premasticate food in order to give their young children lukewarm, chewed food mouth to mouth. In this context, the romantic kissing of couples might be a behavior derived from this type of feeding in childhood and might make adults recall the pleasure they experienced as newborns, especially since— as the well-known sensory homunculus model* reminds us—the mouth is an area of the body that is extremely sensitive to tactile stimuli. In her book *The Science of Kissing: What Our Lips Are Telling Us*,[336] American researcher and science journalist Sheril Kirshenbaum proposed an alternative hypothesis about the origins of this practice. Going back in time, Kirshenbaum tells us that in the age of the first humans, every encounter began with a sniff of the other person just like among other land mammals. These bodily olfactory exchanges likely transformed over time into nose-to-nose and then mouth-to-mouth exchanges, so as to *inhale* a partner from as close as possible. Even so, romantic kisses do not appear universally but are still found in 90% of human cultures.[337] As for this practice's advantages, there are multiple. For the Dutch researcher Rafael Wlodarski at the University of Oxford, kissing has evolved to make it possible to assess an individual's state of health and genetic compatibility by using odor signals. At the beginning of a relationship, a man and woman subconsciously gauge their genetic heritage through their MHC (detected by smell). Therefore, the French kiss might be useful in evaluating a partner in order to ensure the birth of strong offspring.

But once love is well established, why do couples still kiss? To answer this, Wlodarski and his colleagues recruited 902 people from the United States and England to describe their kissing frequency and their relationship satisfaction. The results were crystal

---

* The sensory homunculus is a representation of somesthesia (perception of bodily sensations) found in the cerebral cortex. It is a model of the human body with a mouth, hands, face, and feet that are disproportionately large compared to the rest of the body. This is what we would look like if our anatomy were proportional to the areas of the brain dedicated to sensory processing.

clear! The two are closely linked—the more people are satisfied and happy with their partner, the more they kiss.[338] Beyond its function as a gauge of genetic compatibility, kissing is evidence of the affection that solidifies bonds. It is more frequent just before sexual intercourse, which helps to bond a couple, and is therefore important in establishing a relationship and maintaining it over time. Its role in sexuality, however, was found to be limited. As English psychologists Sandra Murphy and Polly Dalton at Royal Holloway, University of London[339] point out, the reason why we close our eyes during romantic kissing is not to avoid desire but to better concentrate on our tactile, olfactory, and gustatory sensations, as a way to stop paying attention to the visual stimuli that often overpower other sensations.

If love really does exist among nonhumans, then what about friendship? Researchers set out to test the existence of this type of bond in cattle. Some individuals were confined with the individual they interacted with most in their daily life, while others were placed for thirty minutes with an individual they had never met before. Heart rates and cortisol levels (the stress hormone) were recorded throughout the experiment. What happened? In the presence of their *best friend*, the cattle presented stable and regular heart rates, but those placed with a new individual were very stressed.[340] Anecdotally, the American scientist and neuroeconomist Paul Zak at Claremont Graduate University in California became interested in an inseparable male goat and male dog living in a shelter in Arkansas, an odd couple to be friends. He took blood samples from the two animals, put them in the same enclosure, let the two companions interact and play, then took new blood samples fifteen minutes later. The results were astounding. The dog's oxytocin levels increased by 48%, showing his attachment to the goat that he considered, according to the researcher, to be a very good friend. For the goat, the increase was even more significant: 210%. This is a rate typically observed in a human in love, gazing adoringly at their beloved.

A Partner in Life and in Death?

Western societies extol the monogamous couple, in life and in death, as if monogamy were one of the core values defining us as human beings, the very reflection of our essence. But what is it really? Is our monogamy *natural*? Could it rather be the result of our culture, muffling biological determinism? Biologists distinguish social monogamy, which consists of having a partner with whom you spend a lot of time and with whom you raise your young, from sexual monogamy, which does not allow any other sex partners. Most human beings are social monogamists but not necessarily sexual monogamists. Social monogamy is quite common among primates, occurring in about a quarter of the species within all groups of primates and their ancestors. In other mammalian species, however, it appears more disparately and less frequently (estimated at 5%).

In humans, the choice of an exclusive mate evolved directly from an ancient reproductive strategy called polygynandry (a variant of polygamy). In this mating system, multiple males mated with multiple females in a stable way—regularly changing sexual partners but always in the same group—to reproduce. Monogamy in primates is estimated to have appeared several million years BCE, its exact date still the subject of debate. Once it existed, this reproductive strategy appears to have remained fairly stable over time in these species. From time to time, however, some of our ancestors did return to the old polygynous systems. The appearance of social monogamy in human beings likely played a major role in our evolution. In a paper on monogamy, Blake Edgar calls attention to a remark made by Bernard Chapais, an anthropologist at the University of Montreal, pointing out that "pair bonding likely led to human social organization."[341] Christopher Opie, another specialist in human evolution, was a researcher at University College London when he carried out his pioneering work in this area. He was interested in the evolution of 230 species of primates in order to assess the best explanation for the appearance of social monogamy in humans.[342] Three hypotheses emerged. The first was that monogamy had appeared with the increased need for biparental care because the offspring would benefit from being raised by both their male and female parents. This could be crucial for their survival, especially during the birth of twins.[343] The second hypothesis was that mate guarding led to monogamy, which allowed males to no longer compete for females.[344] And the third explained monogamy as a way to reduce the risk of infanticide, since in mammals it is not unusual for a male to kill the young of a lactating female so that the menstrual cycle will resume and another ovulatory cycle will occur, and thus give him the opportunity to reproduce.[345] The study's results revealed that there is indeed an evolutionary correlation between social monogamy, paternal care, female exclusivity, and the reduction of infanticide; but reduced infanticide would have happened before the change in reproductive strategy, whereas the other two explanations only appeared because of it. This means that social monogamy appeared in primates as a *solution* in initially polygynous systems to overcome the high rate of infanticide and thus promote the survival of offspring. In fact, in chimpanzees, who practice polygyny, several cases of infanticide have been reported by primatologists.[346] Scientific observation of chimpanzee births in the wild has shown that, generally, female chimpanzees isolate themselves when giving birth, then raise their young discreetly (a sort of maternity leave), and do not return to the group until later—a strategy probably adopted to limit the risk of the young being killed by the males. Primatologist Adriana Lowe at the University of Kent in the United Kingdom has even shown that females adjust their behavior according to the social ranks of males. When the females have a baby, they associate more readily with dominant individuals and particularly avoid exposing their young to males that occupy lower ranking positions, or those that are unlikely to be the fathers of their offspring.[347]

These studies converge on the hypothesis that our current reproductive strategy is inscribed, at least in part, in our genes. To help us see things more clearly, recent experiments in meadow voles and prairie voles were conducted by an American team of neuroscientists. The polygamous meadow vole leads a solitary life, while the prairie vole is social and monogamous, with faithful couples that raise their young together. In prairie voles, the researchers found a greater number of receptors for a neurotransmitter called "vasopressin," in the nerve center related to pleasure. This compound is secreted especially during sexual activity, and its release allows the male to pair-bond with the female. There is therefore a genetic basis for monogamy in this grassland rodent. In an experiment bordering on science fiction, researchers transferred the "monogamous gene" to the polygamous vole species using viral vector gene transfer. Male meadow voles that were initially polygamous but that received the "monogamous gene" exhibited new prairie-vole-like behaviors, such as pair-bonding with a female and showing her increased attention.[348]

Like voles, are human beings prisoners of their DNA, their proneness for monogamy determined by their genes? In a study on the vasopressin receptor gene (named AVPR1A) in human men, researchers demonstrated that individuals' perceptions of the quality of their romantic relationship were associated with the number of copies of the 334 allele they carried. Men carrying two copies of this allele reported more marital problems compared to those not carrying the allele.[349] In fact, the genetic basis of monogamy does not stop at AVPR1A. Like the vasopressin receptor gene, the oxytocin receptor gene impacts the pair-bond quality described by women. Scientists have found that women carrying the A-allele of this gene are 50% more likely to experience serious marital conflict compared to women not carrying this allele.[350] And what about men? Oxytocin does not just affect women. A team of German psychiatrists at the University of Bonn studied the role that the so-called "love hormone" could play in maintaining monogamous relationships in males.[351] By administering oxytocin intranasally, the scientists demonstrated that it encouraged single men to approach an attractive woman, while men in a monogamous relationship kept their distance. The natural release of oxytocin in males already in a relationship allows them to avoid the *romantic* signals from other women. In this way, oxytocin may help promote fidelity to a romantic partner in humans and other monogamous animals.

Other research has revealed that identical twins (with identical genes) have a fairly similar number of sexual partners as one another during their lifetimes compared to fraternal twins (whose genes are as different as non-twin siblings).[352] These studies underscore the role of genetic determinism in human monogamy. Franck Cézilly, professor of behavioral and evolutionary biology at the University of Burgundy, reminds us, however, that monogamy should not be considered in only natural terms. Human beings are able to adapt their reproductive paradigm more easily than other animals, particularly depending on socioeconomic contexts. The more a cultural model tends toward democracy, the more necessary division of labor becomes, and the more it strengthens monogamy. Conversely, in societies in which roles are more unequal, polygamy allows one to benefit from the wealth of an individual with many resources (even as someone's second spouse), rather than being in an exclusive relationship with a person without resources.

Humans Do Not Have a Monopoly on Pleasure

Why did nature create reproduction? After all, imagine if we had been born as asexual immortal beings without the need to mate. But this would overlook the fact that the world is constantly changing and that all earthly creatures must adapt to survive. Even if we were born into timeless bodies, sooner or later we would be overtaken by environmental changes and eventually die. In that case it would not be from old age but because our ill-adapted bodies would no longer hold up in a new environment. Reproduction is the key that allows new individuals who are better adapted to their environment to come into existence. While genetic mutations within a single generation are sometimes miniscule, they become monumental over the course of many generations. Reproduction allows each species to constantly renew itself and modify its DNA to create beings that are optimized at that exact moment in time (those that are less optimized will perish before the others). This shaping of living things takes place through a subtle balance between death and reproduction. It is the cornerstone of the evolutionary process, which has allowed life to flourish over the past three billion years. The diversity of life forms tends toward an increasing complexity to ensure that life persists, even during extreme climatic shifts.

In humans, when intercourse is consensual, sexual activity is (very often) synonymous with pleasure. In our societies, sexual relations are hidden from the eyes of others. The immense success of pornographic films throughout the world, however, clearly shows the importance sexuality holds for our species, which is very receptive to visual stimuli. While sexual activity decreases with age, some couples maintain very active sex lives. Young or old, human beings often think of sex as having a psychological aspect. Additionally, to make love, we must desire; and in order to desire, we must engage in cognitive processes. Some even claim that there can be no sex without love. Be that as it may, due to both our brain development and culture, cognitive processes have a structural impact on human sexuality, as demonstrated by Serge Wunsch, neuroscience instructor-researcher at the École pratique des hautes études (an elite research and higher education institution) in Paris.[353] This explains how Catholicism's influence on Western societies has created the ideal of chastity, over millennia, with sexuality for the sole purpose of reproduction. On the basis of anthropological, historical, and ethological studies, Wunsch points out that, in humans, "reproductive behavior is made up of fundamental innate and instinctual elements combined with learned skills." He notes a fundamental difference between rodents and hominids in the "structural and functional changes in the nervous system that induce changes in the control of motivational behaviors."[354] Does this mean, however, that unlike us, "beasts" only have sex when driven by their instincts, never separating reproduction and sexuality? If this were true, then when in heat, females would be bombarded with male advances for the sole objective of reproducing.

But that is not the case. In primates, but also in all animals with developed cognitive skills, sex is often a source of pleasure, and sometimes as much for females as it is for males. Some females actively seek sexual interaction (a phenomenon called "proceptivity"); others mate outside of their reproductive period, during pregnancy, during menstruation, or while incubating eggs. This non-procreative sexual activity constitutes a significant proportion of their sexual behavior. This was made clear by the numerous observations made in 1999 by Canadian biologist and sexologist Bruce Bagemihl[355] of several species, including addaxes (screwhorn antelopes), mountain goats, rhesus macaques, proboscis monkeys, golden lion tamarins, and even the sea birds known as guillemots. Through extravagant courtship displays, subtle hormonal shifts that help generate simultaneous desire, and the magic of brain chemistry that sometimes creates unbreakable attachment bonds, evolution has selected countless behaviors that promote reproduction. In other words, human beings are not alone in experiencing pleasure. So does this mean that pleasure has a vital role? As Thierry Lodé, professor and specialist in animal sexuality at the University of Rennes, explains, "For the sexual act to remain in the course of evolution, something had to be added. Pleasure contributes perfectly to it."[356] Blackbirds and chickadees, for example, mate even after eggs have been laid. Against all odds, what motivates an animal to engage in sex is not the perpetuation of the species, but pleasure. This is no doubt the reason why animals display such a wide range of customs. Often the desire felt by a female for a male differs depending on the individual she has in front of her. A plethora of courtship displays thus allow her to choose the most attractive male in her eyes.

Women experience pleasure thanks to a specially dedicated organ called the clitoris. With seven thousand to eight thousand nerve endings, the clitoris is the most sensitive organ in her body. Medical-imaging analyses have shown that it has parts located both inside and outside the body. Two "arms" extend from its base and pass under each of the labia majora. In entirety it measures 9 centimeters (3.5 inches). But the clitoris still remains taboo. The Austrian neurologist Sigmund Freud has his share of responsibility for the feelings of shame we associate with it.[357] His theory of female sexual maturation stated that clitoral orgasm was undignified and that only vaginal orgasm could be considered evidence of mature, fully formed sexuality. Today experts in this field agree that there is no difference between vaginal and clitoral orgasms, both of which are due to the clitoris. According to a 2016 French report on sex education, a quarter of fifteen-year-old girls do not know that they have a clitoris and 83% of thirteen- to fourteen-year-old girls are unaware of its function.[358] The general population, men and women alike, is unaware of this part of the female anatomy. Furthermore, in many of our cultures in which the male sex strives to dominate women, female pleasure is deliberately restricted. According to a UNICEF report published in 2013, female genital mutilation has been performed on more than 125 million girls and women worldwide.[359] The report states that female genital mutilation "is also practiced in pockets of Europe and North America, which, for the last several decades, have been destinations for migrants from countries where the cutting of girls is an age-old tradition." This barbaric and extremely painful practice, which most often is performed without anesthesia, refers to the partial or complete removal of the labia minora, labia majora, and the clitoris, recognized as the most sensitive part of a woman's body. Its goal is to eliminate all female pleasure so as to ensure virginity before marriage, in particular, and subsequent fidelity as a wife. The underlying theory is that after the removal of these organs, because women will no longer be able to feel satisfaction during sex, they will not have the desire to look elsewhere. Beyond serious medical complications and extreme pain, Francesco Bianchi-Demicheli, head of the Psychosomatic Gynecology and Sexual Medicine Office at Geneva University Hospitals, states, "It is often a devastating time, a source of great fear and the origin of real psychological trauma. This can be compounded by a sense of betrayal."[360] It is a strange world humans have created in which male individuals have found no better way to establish their authority than to infringe on the bodily integrity of women.

In animals, the pleasure felt by males can be observed when they cry out or let out deep sighs during orgasm. They probably experience a pleasure similar to that felt by men, with the same cocktail of hormones flooding their brains. And female animals? Do they also experience pleasure? Yes, nature did not leave them out. They experience a phase of excitement before intercourse and therefore even before the possibility of reproduction. For instance, the spotted hyena has a clitoris of such an impressive size that, accompanied by the labia majora, it has an uncanny resemblance to a penis. In fact, biologists often have trouble identifying the sex of these animals. But the hyena is not the only example. By recreating 3D images of the clitoris of eleven deceased female dolphins, researchers Dara N. Orbach and Patricia Brennan at Mount Holyoke College in Massachusetts discovered that dolphins have a clitoral hood where the erectile tissues fuse in a way that is similar to the female human clitoris.[361] With many blood vessels running through it, the clitoris expands during intercourse, offering evidence that female dolphins likely feel intense pleasure during sex. This is also the case for female macaques, which have intense uterine contractions and a sharp increase in their heart rate during orgasm.[362] Even more surprisingly, the frequency of their orgasms increases with the social rank of the male with whom they copulate.[363] Clearly sexual pleasure exists in the animal world, without having a systematic reproductive utility.

Better yet, animals engage in different forms of sexual play, with no reproductive aim. For instance, bonobos engage in oral sex frequently in their youth when playing and exploring,[364] and orangutans engage in it regularly.[365] A Chinese research team led by Libiao Zhang has studied the practice of fellatio in bats in detail and published an explosive article in the prestigious journal *PLoS ONE*.[366] Their incredible results earned them the 2010 Ig Nobel Prize, which is a parody of the Nobel Prize awarded by Harvard University and celebrates the most unusual achievements in scientific research. Among these winged beings, fellatio is not just for discovering another's body, but rather it is an integral part of the sexual *routine*. Beyond providing pleasure, fellatio prolongs the duration of mating, increases the chances of fertilization, and reduces the risk of sexually transmitted infections. Bruce Bagemihl specifies that in animals in general, genital stimulation (including the anal regions) of a partner can also be done using the hands, paws, or fins.[367]

Humans and Animals Enjoy Self-Pleasure

Another sexual practice brings us back to our inner beast and to the impulses over which we have little or no control. It is masturbation. Many consider it to be "against nature" or stigmatize it as a "solitary sin." In the West, until the beginning of the 20th century, most doctors associated this practice with an illness or a perversion.

Why is this sexual practice so upsetting? Islam and Catholicism, for example, condemn it. The Church only tolerates it between spouses if its purpose is to strengthen the couple's unity.[368] And the "beast" emerges quite early in childhood. Masturbation is regularly observed in some infants and even in fetuses. It then reappears in prepubescent children from the age of seven to eight, who have no idea what is happening to their body and who naturally start discovering their genitals by caressing or stimulating them. Many parents do not broach these subjects with their children. Explaining that this behavior is part of the process of discovering one's body is important, however, especially in the development of sexuality. Of course, to preserve everyone's privacy, a child is advised to experiment when alone; but nothing prevents an open discussion, which lets the child avoid feelings of guilt or shame. Child or adolescent, masturbation allows a person to get to know their body better and to prepare for a sexual encounter. In adults, masturbation can complement and enrich an already fulfilling sexuality, or sometimes compensate for sexual frustration. Masturbation should not be considered subsexual. On the contrary, it is an integral part of sexuality. In their book *Petite histoire de la masturbation* (A short history of masturbation),[369] French doctor Pierre Humbert and psychiatrist

Jérôme Palazzolo at Senghor University in Alexandria, Egypt, state that "for young women, masturbation has the function of mastering the ability to achieve orgasm during coital intercourse." For boys, it is also a way of learning to control their phallus. This practice thus promotes access to pleasure and is perfectly complementary to coitus between two partners.

While this behavior seems innate in humans, could it have had an evolutionary origin? We do know that many nonhuman primates, both males and females, perform autosexual behaviors.[370] Frans de Waal recounts this practice in bonobos, stating, "If female bonobos habitually masturbate, however, this activity must surely produce enjoyable sensations."[371] German researcher Ruth Thomsen at the Institute for Zoo and Wildlife Research in Berlin showed in a study that all male macaques on Yakushima, an island in Japan, masturbate without exception,[372] using the same technique as human males. Lower-ranking males spent more time masturbating than higher-ranking males, as the latter had easier access to females. But while animals are capable of self-pleasuring, how do those without hands or fingers do so? Different practices as varied as they are surprising have been observed in a large number of animal species, demonstrating their creativity. Porcupines can rub their genitals on different objects (rocks or pieces of wood), while male horses slap their genitals against their stomach. Walruses use their flippers to caress their genitals, and male dolphins spend a considerable amount of time rubbing their penises. The French cetologist Éric Demay even described the example of Dolphy, an isolated female who used buoy chains like a sex toy for self-pleasure.[373]

Homosexuality in Nature

As we can see, a large majority of species are motivated to seek sexual pleasure through a wide range of practices. So what about homosexuality? Currently, gay people still face discrimination around the world. In Morocco, for example, homosexuality is a crime punishable by imprisonment. In Iran, Sudan, and Saudi Arabia, it brings a death sentence. In Egypt, anal examinations are still performed by some doctors to identify gay men. Where did this idea come from that only a man and a woman have the right to love each other and have sex? According to Genesis, the first book of the Bible, God created man and woman as two complementary beings—without a second Adam or a second Eve. The teaching of the Catholic Church has always described homosexual acts as "intrinsically disordered." Pope John Paul II even listed homosexuality as one of the four great sins of his time, along with contraception, divorce, and euthanasia. But in 2020, Pope Francis broke with doctrine, expressing support for same-sex civil unions, though he remains opposed to same-sex marriage.[374] For Muslims, the Koran states in sura 26, verses 165 to 167, "What! Do you, of all the world's people, approach men [with lust], and leave aside what your Lord has created [and made lawful] for you in your wives? No, indeed! You are a people exceeding all bounds [of decency]. They responded: 'If you do not desist, you will most certainly be cast out [from our land].'" The same prohibition exists in Judaism. In two verses of the Torah (Leviticus 18:22, 20:13), male homosexuality is considered an abomination, though some interpret these verses as a prohibition of incest between men rather than of homosexuality.

In fact, in the eyes of monotheistic religions, sexual practices that do not lead to procreation are generally regarded negatively. Considered immoral, they bring us back to our impulses and therefore to our damned animal nature. Only sexual practices for procreative purposes between a married man and woman are considered "moral." Has this always been the case? Let's look at the sexual practices of our predecessors. And to begin our journey into the past, we must start by reopening the doors to prehistory—reopening because, despite the incredible paleontological research of recent centuries, until a few decades ago, a variety of taboos on prehistoric sexuality prevailed, which had prompted scientists to share sanitized descriptions of hominids' sexual practices. In 2017 scientists Javier Angulo and Marcos García-Diez broke the code of silence by presenting an exhibition entitled *Sexo en piedra* (Sex in stone) of pieces found near the site of Atapuerca in Spain. A real coming-out for the Paleolithic Age! Among these archaeological marvels, a fourteen-thousand-year-old stone engraving illustrates a scene of sodomy between two men, while on a twelve-thousand-year-old plaque we can discern two women sensually intertwined.[375] If we catapult ourselves a few millennia later to ancient Greece, the concepts of heterosexuality and homosexuality simply did not exist, as bisexuality was the norm. In this ancient Greek world, desire could be expressed equally toward men or women; only aestheticism prevailed. It was acceptable to find any individual attractive, regardless of gender or even age. The only norm in this respect was that the man who penetrated his partner was always considered dominant, regardless of the sex of his lover. The Greek practice of pederasty was a rite of passage for adolescent males, a sort of "education of youth," which today seems quite shocking. With its own complex etiquette, an adult man called the *erastês* played the role of teacher to an adolescent called the *erômenos*, who had to submit to him. Pederasty, the "love of boys," is documented in the works of the greatest philosophers, from Plato to Plutarch.

In ancient Rome, before life was abruptly interrupted on August 24 and 25 in the year 79 CE when Pompeii was buried by Mount Vesuvius's volcanic eruptions, all visitors to the city could admire the erotic art of its magnificent wall frescoes. In a series of paintings in the Suburban Baths, several scenes depict a variety of sexual practices, including oral sex and homosexual activities between men and between women—and sometimes in very acrobatic positions. As the

Italian archaeologist Antonio Varone, director of the excavations of the ancient flourishing city on the Gulf of Naples, points out, "Pompeii is not the city of sin, the city of perversion and lust, a new Sodom whose divine punishment would have been well deserved,"[376] because the Romans were able to maintain a relationship with sex that was free from prudishness and sin. Julius Caesar (100 to 44 BCE) himself was given the nickname "every woman's man and every man's woman"[377] by Curio, who held the position of plebeian tribune. The moniker was due to the relationship Caesar allegedly had with Nicomedes, the king of Bithynia. Later, Roman emperor Hadrian (76 to 138 BCE) fell in love with a Greek youth named Antinous. After his lover drowned in the Nile, Hadrian erected a temple and a city in Antinous's memory. The cult of Antinous continued for almost two hundred years throughout the empire. It was not until the arrival of Christianity that sexual acts between same-sex partners were condemned. Ancient Rome's ten centuries of sexual freedom came to an end when the emperor Theodosius, a Catholic convert, issued an edict on August 6, 390, condemning "passives"—or "men who copulated like women"—to death. In the Middle Ages and until the French monarchy was abolished, homosexuality was a crime against the natural order established by God and carried the death penalty in most European states. In 1791 France decriminalized these acts, but gay people continued to be persecuted for a long time thereafter, particularly during World War II when they were deported to concentration camps solely because of their sexual orientation.

In science, homosexual behavior is differentiated from homosexuality. Therefore, just because we observe homosexual behavior in members of a species, this does not necessarily indicate homosexuality in those individuals, in the sense of an exclusive pair attachment between two partners of the same sex. And, as Thierry

Lodé explains, "all sexual behaviors exist in nature." In Greek antiquity, Aristotle mentioned in his *History of Animals* that erotic or sexual acts between partners of the same sex exist in quails, partridges, and roosters. At the time, the philosopher put forward a hypothesis to explain these behaviors, stating, "Homosexual contact between males in partridges, quails, and roosters are the result of dominance behaviors. The victors of fights mount their defeated rivals." Many believed homosexuality was a second choice for animals that heterosexuality escaped. Since then, other explanatory hypotheses have emerged. The study of homosexual behaviors in animals gained momentum in the 1990s, and the work of ethologists made it possible to break with the reductionist understanding of these behaviors (attributing a purely reproductive function to sexual behavior) in favor of an understanding that animal behavior is variable.

Contrary to the belief that any act of copulation in animals is geared toward reproduction, scientists have discovered that sexual acts can have other functions (social cohesion in particular) and, as we have seen, that seeking sexual pleasure can generate by itself a wide range of behaviors. On closer inspection, researcher Bruce Bagemihl found bisexual relationships in nearly 450 animal species.[378] Thierry Lodé also highlighted the existence of bisexual behavior in a large number of mammals, including lions and many species of apes.[379] After studying 93 species of birds, biologist Geoff MacFarlane at the University of Newcastle in Australia revealed that 5% of their sexual activities are homosexual.[380] Partners of the same sex can *mount* each other and engage in intercourse with genital contact. Interestingly, MacFarlane noted that homosexual behavior in males was more common in polygamous species, while homosexual behavior in females was found more frequently in monogamous species. In an article about his research published on the Live Science site, it was explained that the lighter an animal's parental duties, the more it would mate with multiple partners, regardless of their sex.[381]

In fact, many species have a much broader sexuality than previously thought. Zebras also have homosexual relationships, as do elephant seals, chameleons, and bonobos. Among bonobos, whose expansive sexuality is widely commented on, both males and females frequently engage in relations with partners of the same sex. Not just for obtaining pleasure, sexuality also helps to ease tensions within the social group. In gorillas, though much less sexually inclined than their bonobo cousins, scientists have observed lesbianism. In his evocatively titled book *Biological Exuberance: Animal Homosexuality and Natural Diversity*, Bruce Bagemihl observed male dolphins regularly rubbing their penises against one another[382] as a way, he believed, of forming strong bonds with other males who traveled together and defended themselves against predators, one being able to rest while the other kept watch. Bagemihl recounted a similar phenomenon observed in killer whales during which they reunite in groups exclusively comprised of males, after leaving their matrilineal group. In pairs or sometimes in groups, the killer whales have a grand time rolling over each other on the water's surface, splashing and rubbing their bodies against each other. Orgasms have been physiologically observed during homosexual intercourse among animals, including in female macaques.[383] But biologists, in keeping with the logic of natural selection, were slow to admit that the goal of animal sexuality was not solely reproductive, continuing for decades to seek justifications for homosexual relations between animals. Confronted with the ever more abundant discoveries of ethologists, the scientific community had to change its tune. Like humans, animals engage in intense carnal pleasure, where heterosexuality is far from being the norm.

Therefore, in the animal world, homosexuality exists beyond *only* homosexual behavior since it occurs at several levels of a relationship, including during courtship, pair bonding, sexual relations, life as a couple, and even parenting. At Bremerhaven Zoo in Germany, six male Humboldt penguins formed three same-sex couples and *adopted* pebbles as eggs. Despite the introduction of female penguins into their enclosure in order to promote reproduction, the homosexual couples maintained their relationships. This observation is of course biased by captivity, but the fact that the penguins were not interested in the introduced females shows that in these animals, the formation of same-sex pairs creates lasting bonds. Two of these homosexual penguins later became the adoptive parents of a baby penguin that they hatched themselves. The egg had been rejected by its biological parents before hatching and had been offered to the three couples. One of the homosexual couples decided to take care of it. This story appeared in the French newspaper *Le Monde* in an article reporting that the penguins' trainer said, "The two penguins welcomed the egg with joy and incubated it proudly." Once the egg hatched, the couple behaved with their chick just like heterosexual parents do, grooming him and feeding him fish mash.[384] Therefore, homosexuality is not a secondary sexual activity since, as we can see, these gay animal couples raise young (conceived with the help of external partners) and remain together for many years. In the natural environment, certain species of lizards that reproduce by parthenogenesis (a mode of reproduction in females based on self-fertilization) exhibit exclusive homosexuality between females (the males having disappeared), which is necessary for ovulation. Once again, Geoff MacFarlane and his team have shown that in greylag geese 20% of pairs are formed by animals of the same sex.[385] In a study of Laysan albatross conducted by ecologist Lindsay Young at the University of Hawaii and her collaborators, the proportion of lesbian couples on Oahu was estimated to be 31%.[386] Rather than "kings of the sky," Baudelaire should have written of "queens" in his poem "The Albatross"[387] because the population of these seabirds included more females than males. While the albatross population was in decline, these birds managed to maintain an adequate reproductive rate through the formation of female-female pairs. Lesbian mothers raised fewer chicks than heterosexuals, but their efforts have repopulated the island.

While nature can be used to explain human behavior from a biological and evolutionary point of view, we must be careful, however, not to give what is "natural" a moral standing. A plethora of situations that exist in nature, such as forced coitus and infanticide, do not make these behaviors acceptable. Moreover, evolution is constantly shaping behaviors. For a given species, what was "natural" in the past is not necessarily so today (and vice versa). Nature is not set in stone, and neither are behaviors. In the human species, as long as it occurs with the informed consent of both partners, any sexual behavior is a fundamental right and a matter of freedom. In my view, sexual orientation is therefore not a moral standard to be defined. It is a right to love freely.

Do Animals Behave Immorally?

Evolution tends toward neither good nor evil, but toward the proliferation of organisms and behaviors. In many animals, the sexual act, often synonymous with shared pleasure, can also prove to be very violent. Within polygamous species, males can be very aggressive toward each other and toward females. For instance, even though known for his calm nature, the male orangutan can harass a female who refuses his advances, even forcing copulation,[388] which in human terms we would call rape. In animals, ethologists prefer to use the term "sexual coercion," because to date no study has evaluated the emotions felt by individuals who are harassed or who have been subjected to unwanted coitus. The coerced sexual act experienced by a female orangutan, however, appears to be devoid of pleasure. Here arises the essential question about whether the "rapist" was aware of the pain caused for the "forced" subject. Can we incriminate an animal if it is not aware of having acted *badly*? Even though most animals have a protomorality that dictates their rules of conduct, they do not have group consensus or a system of justice that allows them to punish cruel individuals. No aggressive male fears punishment for "inappropriate" behavior, and no individual is likely to be aware of engaging in *deviant* sexual behavior.

In their book *A Natural History of Rape*, biologist Randy Thornhill and anthropologist Craig Palmer indicate that forced sexual intercourse in humans is the result of Darwinian natural selection.[389] According to them, rape can be explained by natural impulses and has an evolutionary origin. Being a biologist, I understand this argument to the extent that the periods when our ancestors were completely polygynous were probably characterized by a high rate of aggression in males, who certainly raped the females of their groups regularly. Justifying the existence of rape by its practice in ancestral times, however, amounts to omitting the thousands of years of evolution and culture that separate our ancient polygamous systems from our contemporary societies, as well as forgetting the appearance of social monogamy. According to the ethologist Frans de Waal, for rape to have been selected by evolution as a key process of our reproduction, the rape act would have had to allow a higher rate of pregnancy in women and therefore confer a reproductive advantage to rapists.[390] This is, of course, not the case. Rape is not a winning reproductive strategy. If it had been a driving mechanism of reproduction, social monogamy would not have been so successful throughout the world. Why couple up if it has no adaptive advantage? Joseph Henrich, a psychologist and economist, and his colleagues at the University of British Columbia in Canada also point out that sexual assaults decrease drastically in monogamous systems compared to polygamous societies.[391]

In animals, even more surprising practices have been brought to light, such as mating attempts between different species. In theory, a plethora of obstacles generally inhibit these behaviors, including anatomical incompatibilities, the fact that sexual attraction only has an effect on individuals of the same species, and "molecular securities" that prevent fertilization since the ovum and sperm must recognize each other at the membrane level so that the sperm can penetrate the ovum's membrane. But in 1995 on the Japanese island of Yakushima, the photographer Alexandre Bonnefoy immortalized the image of a macaque mounting the back of a sika doe in an attempt to mate. The doubtful doe did not seem to appreciate the primate's assault. The most detailed example, reported in an article in the journal *Polar Biology* by a team at the University of Pretoria in South Africa, concerns the mating of male Antarctic fur seals with king penguins.[392] The researchers specified in this case that there was penetration. A photograph even shows one of the penguins bleeding profusely just after the forced copulation. The union had no reproductive purpose. So how can these surprising mating attempts be explained? The most likely hypothesis is that male individuals unable to satisfy their needs become frustrated, and their impulses encourage them to direct their sexual behavior toward other animals "crossing their path." An opportunism that is immaterial to the other participant so long as there is neither violence nor penetration. In the case of the fur seals attacking the penguins, this abusive behavior was repeated several times, most likely because the pinnipeds understood that these seabirds were easy subjects to capture and hold under their weight. Given the genital lesions caused by the forced coitus, the penguins certainly experienced significant pain. But the fur seals were not aware of the discomfort or pain of their "partners." They did not feel any empathy toward the penguins, which are their occasional prey (predators categorize their prey very early in life and feel no emotion toward them). The attacking animal does not *desire* the attacked animal; rather the attacker uses the other animal as a substitute object in the same way a human being would use a sex toy.

These *deviant* sexual behaviors between species that are genetically very distant from one another are very different from the *consensual* relations between two animals of different species that have a strong genetic similarity. In fact, it happens that individuals with a relatively close common ancestor, but whom nature has separated and caused to evolve in very different geographical areas, are reunited. Often this happens because of an environmental change that encourages them to move to new areas, or sometimes it happens because they find themselves artificially enclosed in the same captive environment while they never would have met in nature. From these unions are born creatures that one would think came out of Greek mythology, called "hybrids." Mating can occur, but most of the time it results in young that are incapable of having their own offspring. The best-known case is that of the female horse and the male donkey whose young—the mule—will be sterile, just like the infertile offspring of the male horse and the female donkey called a "hinny." In zoos, a tigress and a lion can mate to produce a strange offspring called the "liger." This genetic curiosity of gigantic size and weight (because the growth inhibitor genes normally expressed in the belly of the tigress are no longer expressed) would have a hard time surviving in the wild, but the females born from these unions are fertile while the males are not. As for the "tigon," born of the union of a lioness and a tiger, it does not exceed the size of its parents and has the physical characteristics of its two parents. Sometimes this type of mating takes place in the wild. In 1990, on a small island near Disko Bay in Greenland, scientist and explorer

Mads Peter Heide-Jørgensen noticed an enormous skull on the top of a hunter's tent, not seeming to belong to any species that this whale specialist knew. The hunter then told him that the fascinating creature he had killed had the flippers of a beluga but the tail and gray skin of a narwhal (the famous unicorn of the sea with a fascinating helical tusk). The biologist convinced the hunter to offer this skull to the Natural History Museum of Denmark in Copenhagen. At the time, the scientific community was perplexed by this discovery and suspected the existence of a hybrid, particularly because of the animal's unusual teeth that were a cross between those of the narwhal and those of the beluga. Doubt lingered for about thirty years, however, until experts in genetic analysis managed to determine with certainty that the strange skull was indeed the result of the unbelievable union of a narwhal mother and a beluga father.[393] While the vast majority of crossbreeding between neighboring species results in sterile offspring or young that do not survive, from time to time, the opposite happens and a new species is created that sometimes even replaces one of the two parent species. Hybrids have a massive impact on evolution.

Human beings are a hybrid. Chris Thomas, a conservation biologist at the University of York in the United Kingdom, explains that "genes jump everywhere." He says, "Molecular genetics finds that hybridization between species is more common than previously thought. Darwin spoke of a tree of life, with species branching and splitting. But we're finding it's more of a network, with genes moving between closely related branches as related species interbreed. This hybridization quickly opens up opportunities for evolution."[394] Since hybridizations between neighboring species are possible, the Russian researcher Ilia Ivanov spent his life trying to create a human-chimpanzee hybrid. But the artificial inseminations of female chimpanzees with human sperm were unsuccessful. His research was very controversial, and he was ultimately condemned to exile. In 2019 the Spaniard Juan Carlos Izpisúa Belmonte at the Salk Institute for Biological Studies in California created a chimeric embryo composed of chimpanzee and human cells, but as required by bioethics, it was destroyed before it reached fourteen days old.[395]

What about the human practice of zoophilia? Originally, the word "zoophile" was platonic, devoid of any sexual implications. It meant "who loves animals" (Victor Hugo thus titled his animal protection journal Le zoophile). The term evolved to mean a human's sexual attraction to an animal, even when a human partner is available to them—and that clarification is important. This practice is more widespread than we might think. According to Benoît Thomé, president of Animal Cross, the French association for animal protection, zoophilic pornography sites receive 1.6 million monthly visits and have 150,000 followers in France alone.[396] The Judeo-Christian tradition condemns the "sin of bestiality," considered to be "against nature," not out of empathy for animals but because it lowers humans to what is considered to be most vile: their animality. French law considers bestiality an act of cruelty to animals and has condemned it since 2004. Inflicting serious abuse or abuse of a sexual nature, or committing an act of cruelty on a domestic, tamed, or captive animal, is punishable by two years of imprisonment and a fine of 30,000 euros. But the French position is in the minority. Romania and Finland, for example, have no laws against such practices, and Germany only joined France in banning bestiality in July 2013. Germany has long remained very tolerant of this practice. The French veterinarian Marjolaine Baron, in her thesis entitled "La zoophilie dans la société: Quel rôle le vétérinaire peut-il tenir dans sa répression?" (Zoophilia in society: What role can the veterinarian play in its prevention?),"[397] recounts the story of love at first sight for Michael Kiok, a German librarian and founder of the zoophile association called the Zeta Organization, when he encountered an elephant one day at a circus. Kiok said, "She looked me in the eyes and I immediately saw that she was a charismatic woman." Kiok has a dog named Cessy, with whom he has sex several times a week, and two cats. He helped organize a large annual get-together in northern Germany, where hundreds of people would engage in sex with pigs, horses, sheep, and cows. In addition to the risk of sexually transmitted diseases in humans and the lesions caused to the animals' genitals, such acts, of course, create moral questions. While the "zoosadists" take pleasure in torturing animals during coitus, the "classic" zoophiles consider their sexual practices to be true demonstrations of love and refute any idea of mistreatment. They claim a freedom from morals and a "natural" sexual orientation. The Australian philosopher Peter Singer, known for his stance in favor of animal rights, has a surprising view: he believes that zoophilia should be tolerated if no harm is done to the animals. According to him, aversion to zoophilia is the result of irrational speciesism and our anthropocentrism.

But what is the difference between the mating of animals of different species and zoophilia? The animals that engage in these inter-species "rapes" do not *desire* the partner they are assaulting. In other words, the male fur seal is not attracted to the king penguin, nor is it demonstrating its affection by penetrating the penguin by force; rather it is using the penguin as an object. The zoophilic human being expresses a genuine preference for an animal deemed attractive, even when human partners are available. In my opinion, this constitutes a fundamental difference. Zoophilia is a human mental construct and therefore a mental disorder, whereas attempts at interspecific mating between animals are the result of opportunism *for lack of a better option*. This point is crucial because it instantly sweeps away the zoophile defense of naturalness. This practice cannot in any way fall within the framework of sexual freedom, since unlike the fur seal and the penguin, humans are perfectly aware that an animal that cannot speak is not able to give its unequivocal consent. And sex without consent is sexual abuse. Even if an animal seeks out a sexual interaction with a human (for lack of a better option), a man or a woman equipped with a *normally* functioning brain has the power to suppress his or her impulses thanks to moral standards, while the animal—not having constructed rules of collective morality—is not able to do so. Finally, because humans' destructive and degrading capacities are so alarming, it is our responsibility more than any other animal's to respect the differences of nonhumans and not burden them with our own vices.

Reconciling with Our Animality

The worst form of absurdity is to accept the world as it is, and not fight for the world as it should be.

JACQUES BREL

For millennia, humans have built belief systems to ward off their fear of death. By developing metaphysical thought, they succeeded in creating an *otherworld* and allaying their existential fears. Their first religions were rooted in an animist framework. As part of a whole, *H. sapiens* granted a special place to animals, as demonstrated by the overrepresentation of animals in cave art. *H. sapiens'* interactions with nature were balanced, based on reciprocity. When the first deities emerged, endowed with zoomorphic attributes, animals continued to play a central role.

A shift occurred when our ancestors felt the need to identify themselves with their gods, an anthropomorphization fraught with consequences. For the first time, humans were no longer a single piece of a bigger puzzle but became the ones for whom the puzzle was created. Animals were stripped of their sacred nature, and sacredness only remained in the human being and the divine. From this new conception of the world emerged a dualism between human and beast. And the change was radical. To become closer to the divine, *H. sapiens* thought they had to shed their animal status. Nature and nonhuman creatures were now nothing more than a pool of resources to draw from. And anything that brought *H. sapiens* back to animality was repulsive. They had to bury the beast deep within them, denying its existence, so they could rise toward the heavens. *H. sapiens* no longer owed the world anything; the world now owed them everything!

In 1859 Charles Darwin and his theory of evolution shattered this model of thought by reintegrating humans into the animal kingdom. But reconnecting with one's animality was not an easy task. How could humans accept the idea of being relegated to the status of beast, even though they had been striving to distinguish themselves from it for thousands of years? It was a very hard pill to swallow. It was no longer a question of dithering around philosophical questions or theological positions, however, since science, impartial and objective, revealed and continues to reveal discoveries on a daily basis that cannot be ignored. Our behaviors and our skills, however emancipating, can be understood in the context of our animality. But then, if human beings are *just* animals, even if unique ones, what does that leave us for delineating ourselves? Since we construct our *human* identity through culture and how we want to be seen, it is our actions and values that give meaning to our passage on Earth. Reconciling with the "beast within" not only allows us to know ourselves better but to be open to a new worldview in which animality is no longer derogatory. Wonder awaits anyone who dares to lift their blinders, because the kingdom of beasts—these masses thought to be devoid of intelligence and feelings—is revealed to be a universe of genius, vibrating with emotion. But it is also a violent realization; reestablishing our connection with our animality and with animals forces us to take stock of the consequences that our long disconnection with them has had, and of the urgent need to remedy it. Poached, transported like merchandise, confined in spaces that do not respect their needs, slaughtered without being stunned first; there are billions of sentient beings who suffer because of our anthropocentric way of thinking. How long will we continue to exploit animals so we can convince ourselves that we are not also animals? This is the challenge of our time: either we regard other forms of life with humility and respect, or we blindly continue in our occult beliefs, which reinforce our place in the universe but condemn thousands of species to extinction and precipitate our own demise in the process.

The ubiquity of meat in our societies should also make us question ourselves. We no longer see the connection between the steak on our plate and the creature from which it came. We cast aside the disturbing thoughts of how it was killed and ease our conscience by rationalizing that the animal is there to serve us. Yet, if slaughterhouses had transparent walls, wouldn't the whole world become vegetarian? In her book *L'humanité carnivore* (Carnivorous humanity),[398] the French philosopher Florence Burgat, research director at the French National Institute of Agricultural Research (INRA), explains how the "superstructure" represented by our meat diet "aims to normalize a certain type of relationship with animals, which in turn defines humanity (and animality) in terms of the division between those who eat and those who are eaten." She suggests that beyond the culinary attachment to meat, humanity may especially be holding on to its murderous relationship with animals as a mark of its domination.

But then how do we reconcile with the "beast within"? First of all, we need to free ourselves from this way of thinking that separates humans from nature and elevates us to the center of the universe, the masters of the earth, the ultimate goal of evolution. Then, we need to adopt a critical view of all the preconditions this dualism established. Animals are pests, or for the circus, or for consuming only in our imaginations! But there are many ways of existing in the world, with no more nobility in being a human, a wolf, or an earthworm. Animals are not here to serve us; they lead their own existence. So we must reinvent humanity by incorporating non-anthropocentric ethical imperatives that respect the diversity of life forms into a new consciousness that broadens our empathy toward all creatures with whom we share our existence.

This will require a better knowledge of ethology. But rather than looking at animals through our own eyes, we must learn to look through theirs. This exciting change of perspective (which the Animal Worlds series offers) opens the doors to unexpected worlds, encourages us to reimagine humans as part of a whole, and invites us to show new consideration for nonhumans.

ACKNOWLEDGMENTS

To Professor Bertrand Deputte, ethologist, my mentor and friend, who reviewed the content of this book.

To my husband, my daughters, my parents, my sisters, and my friends for their encouragement and love.

To Olivia Recasens, Joanna Blin, Camille Couture, and the entire humenSciences team for their kind support.

To all the animals who have shared or who currently share my life and who are definitely not "beasts"!

NOTES

1. Denying Our Animality

1. Charles Darwin, *The Life and Letters of Charles Darwin, Including an Autobiographical Chapter*, ed. by his son Francis Darwin (London: John Murray, 1887).

2. Charles Darwin, *On the Origin of Species by Means of Natural Selection, or the Preservation of Favoured Races in the Struggle for Life* (London: John Murray, 1859).

3. Desmond Morris, *The Naked Ape: A Zoologist's Study of the Human Animal* (London: Jonathan Cape, 1967).

4. Carina M. Schlebusch et al., "Human adaptation to arsenic-rich environments," *Molecular Biology and Evolution* 32, no. 6 (June 2015): 1544–1555.

5. Pascal Picq, "L'humain à l'aube de l'humanité," in Pascal Picq, Michel Serres, and Jean-Didier Vincent, *Qu'est-ce que l'humain?*, (Paris: Le Pommier and Universcience, 2010), 33.

6. Debate between Pascal Picq and Father Laurent Stalla-Bourdillon recorded by France Culture, December 2019, at the Collège des Bernardins, Paris.

7. Ewen Callaway, "Mystery humans spiced up ancients' sex lives," *Nature* November 19, 2013.

8. Kerri Smith, "Modern speech gene found in Neanderthals," *Nature,* October 18, 2007.

9. Ruggero D'Anastasio et al., "Micro-biomechanics of the Kebara 2 hyoid and its implications for speech in Neanderthals," *PLoS ONE* 8, no. 12 (December 18, 2013).

10. Bruce L. Hardy et al., "Direct evidence of Neanderthal fibre technology and its cognitive and behavioral implications," *Scientific Reports* 10 (April 9, 2020): 1–9.

11. D. L. Hoffmann et al., "U-Th dating of carbonate crusts reveals Neandertal origin of Iberian cave art," *Science* 359, no. 6378 (February 23, 2018): 912–915.

12. Philipp Gunz et al., "Brain development after birth differs between Neanderthals and modern humans," *Current Biology* 20, no. 21 (November 9, 2010): R921–R922.

13. Marcia S. Ponce de León et al., "Brain development is similar in Neanderthals and modern humans," *Current Biology* 26, no. 14 (July 25, 2016): R665–R666.

14. Takanori Kochiyama et al., "Reconstructing the Neanderthal brain using computational anatomy," *Scientific Reports* 8 (April 26, 2018): 1–9.

15. Vincent Bordenave, "Néandertal était-il trop stupide pour survivre?," *Le Figaro*, April 27, 2018.

16. Allan Hall and Fiona Macrae, "Wanted: 'Adventurous woman' to give birth to Neanderthal man—Harvard professor seeks mother for cloned cave baby," *Daily Mail*, January 20, 2013.

17. Bruno Maureille et al., "The challenges of identifying partially digested human teeth: First description of Neandertal remains from the Mousterian site of Marillac (Marillac-le-Franc, Charente, France) and implications for palaeoanthropological research," *PALEO* 28 (December 2017): 201–212.

18. Christoph Wissing et al., "Isotopic evidence for dietary ecology of late Neandertals in North-Western Europe," *Quaternary International* 411 (August 8, 2016): 327–345.

19. Silvana Condemi et al., "Possible interbreeding in Late Italian Neanderthals? New data from the Mezzena jaw (Monti Lessini, Verona, Italy)," *PLoS ONE* 8, no. 3 (March 27, 2013).

20. Paul Molga, "L'extinction de Neandertal reste un mystère," *Les Échos*, October 25, 2010.

21. Pierre Jouventin, *Kamala, une louve dans ma famille* (Paris: Flammarion, 2012).

22. J. Guichard and G. Guichard, "La naissance de l'art en Europe," in *La naissance de l'art en Europe/Elnacimiento del arte en Europa* (Paris: Union Latine, 1992), 18–28.

23. Iégor Reznikoof and Michel Dauvois, "La dimension sonore des grottes ornées," *Bulletin de la Société préhistorique française* 85, no. 8 (1988): 238–246.

24. Jean Clottes and David Lewis-Williams, *Les chamanes de la préhistoire: Transe et magie dans les grottes ornées* (Paris: Le Seuil, 1996).

25. Jessica Serra, *Dans la tête d'un chat* (Paris: humenSciences, 2020).

26. Alexandra Touzeau et al., "Diet of ancient Egyptians inferred from stable isotope systematics," *Journal of Archaeological Science* 46 (June 2014): 114–124.

27. Isabelle Ohman, "Animaux et rites initiatiques," *Revue Acropolis*, November 2012.

28. Maryse Waegeman, "Plutarque, *Sur l'usage des viandes*," in *L'animal dans l'alimentation humaine: Les critères de choix*, ed. Liliane Bodson, proceedings of the international colloquium of Liège, November 26–29, 1986, published with the assistance of the National Fund for Scientific Research, 1988.

29. Aristotle, *Politics*, trans. H. Rackham (Cambridge, MA: Harvard University Press, 1932).

30. Sophie Madeleine, *Le théâtre de Pompée à Rome: Restitution de l'architecture et des systèmes mécaniques* (Caen: Presses universitaires de Caen, 2014).

31. Pliny the Elder, *Natural History*, trans. John Bostock and Henry T. Riley (London: Henry G. Bohn, 1855), Book VIII, verse 77.

32. Thibault Isabel, *À bout de souffle: Études et entretiens sur l'épuisement du monde civilisé* (Paris: La Méduse, 2012).

33. Conference organized by Yves Modéran (professor of Roman history), "La conversion de Constantin et la christianisation de l'Empire romain," Association des professeurs d'histoire et de géographie (APHG), Caen, June 2001.

34. Georges Ville, "Les jeux des gladiateurs dans l'Empire chrétien," *Mélanges de l'École française de Rome* 72 (1960): 273–335.

35. Michel Pastoureau, *Une histoire symbolique du Moyen Âge occidental* (Paris: Le Seuil, 2015).

36. Hélène Combis, "Truie condamnée à mort, dauphins exorcisés . . . les étranges procès d'animaux au Moyen Âge," *France Culture*, December 28, 2018.

37. Michel de Montaigne, "Apologie de Raimond Sebond," *Essais*, book 2, chapter 12 (Paris, 1580).

38. Voltaire, *Il faut prendre parti* (1772).

39. Jean-Jacques Rousseau, *Discours sur l'origine et les fondements de l'inégalité parmi les hommes* (Amsterdam: Marc Michel Rey, 1755).

40. Immanuel Kant, *Leçons d'éthique (1775–1780)* (Paris: Livre de Poche, 1997).

41. Immanuel Kant, *Kritik der Urteilskraft* (Berlin: Verlag Lagarde und Friedrich, 1790).

42. François Boissel, *Le Catéchisme du genre humain . . . , précédé d'un Discours sur les causes de la division, de l'esclavage et de la destruction des hommes les uns par les autres . . . avec deux Adresses . . . à la nation française . . . et avec quelques opuscules relatifs au nouvel ordre de choses* (Milan: Galli Thierry, 1792).

43. Victor Hugo, "Liberté!," *La légende des siècles* (Brussels: Edition Hetzel, Meline, Cans et C, 1859).

2. Intelligence of Their Own?

44. Aurélien Miralles, Michel Raymond, and Guillaume Lecointre, "Empathy and compassion toward other species decrease with evolutionary divergence time," *Scientific Reports* 9 (December 20, 2019): 1–8.

45. Durga Chapagain et al., "Aging of attentiveness in border collies and other pet dog breeds: The protective benefits of lifelong training," *Frontiers in Aging Neuroscience* 9 (April 20, 2017): 100.

46. Jakob von Uexküll, *Streifzüge durch die Umwelten von Tieren und Menschen* (Berlin: Verlag von Julius Springer, 1934); *Bedeutungslehre* (Leipzig: Verlag von J. A. Barth, 1940).

47. Salimbene di Adam, *Cronica, Nuova ed. critica*, ed. Giuseppe Scalia (Bari: Laterza, 1966).

48. Roger Shattuck, *Forbidden Knowledge: From Prometheus to Pornography* (New York: St. Martin's Press, 1996).

49. Natacha Grenat, *Douloureux secret des enfants sauvages* (Paris: La Compagnie Littéraire, 2007), 102.

50. Jean Itard, *Mémoire sur les premiers développements de Victor de l'Aveyron* (Paris: Goujon, 1801).

51. Harry F. Harlow, "The nature of love," *American Psychologist* 13, no. 12 (1958): 673–685.

52. Sonia Harmand et al., "3.3-million-year-old stone tools from Lomekwi, 3: West Turkana, Kenya," *Nature* 521 (May 20, 2015): 310–315.

53. Dora Biro et al., "Cultural innovation and transmission of tool use in wild chimpanzees: Evidence from field experiments," *Animal Cognition* 6 (December 2003): 213–223.

54. Rikako Tonooka, "Leaf-folding behavior for drinking water by wild chimpanzees (Pan troglodytes verus) at Bossou, Guinea," *Animal Cognition* 4 (November 2001): 325–334.

55. Kimberley J. Hockings et al., "Tools to tipple: Ethanol ingestion by wild chimpanzees using leaf-sponges," *Royal Society Open Science* 2, no. 6 (June 1, 2015): 150150.

56. Shinya Yamamoto et al., "Invention and modification of a new tool use behavior: Ant-fishing in trees by a wild chimpanzee (*Pan troglodytes verus*) at Bossou, Guinea," *American Journal of Primatology* 70, no. 7 (July 2008): 699–702.

57. Christophe Boesch and Hedwige Boesch, "Optimisation of nut-cracking with natural hammers by wild chimpanzees," *Behaviour* 83, no. 3–4 (1983): 265–286.

58. Dora Biro, Cláudia Sousa, and Tetsuro Matsuzawa, "Ontogeny and cultural propagation of tool use by wild chimpanzees at Bossou, Guinea: Case studies in nut cracking and leaf folding," in *Cognitive Development in Chimpanzees*, eds. Tetsuro Matsuzawa, Masaki Tomonaga, and Masayuki Tanaka (Tokyo: Springer, 2006), 476–508.

59. Christophe Boesch et al., "Chimpanzees routinely fish for algae with tools during the dry season in Bakoun, Guinea," *American Journal of Primatology* 79, no. 3 (March 2017): 1–7.

60. Jill D. Pruetz and Paco Bertolani, "Savanna chimpanzees, *Pan troglodytes verus*, hunt with tools," *Current Biology* 17, no. 5 (March 6, 2007): 412–417.

61. Adam van Casteren et al., "Nest-building orangutans demonstrate engineering know-how to produce safe, comfortable beds," *Proceedings of the National Academy of Sciences of the United States of America (PNAS)* 109, no. 18 (May 1, 2012): 6873–6877.

62. Benjamin L. Hart et al., "Cognitive behaviour in Asian elephants: Use and modification of branches for fly switching," *Animal Behaviour* 62, no. 5 (November 2001): 839–847.

63. Preston Foerder et al., "Insightful problem solving in an Asian elephant," *PLoS ONE* 6, no. 8 (August 18, 2011).

64. Meredith Root-Bernstein et al., "Context-specific tool use by *Sus cebifrons*," *Mammalian Biology* 98 (September 2019): 102–110.

65. Christian Rutz et al., "Video cameras on wild birds," *Science* 318, no. 5851 (November 2, 2007): 765.

66. Gavin R. Hunt, "Manufacture and use of hook-tools by New Caledonian crows," *Nature* 379 (January 18, 1996): 249–251.

67. A. M. P. von Bayern et al., "Compound tool construction by New Caledonian crows," *Scientific Reports* 8 (October 24, 2018): 15676.

68. A. M. I. Auersperg et al., "Social transmission of tool use and tool manufacture in Goffin cockatoos (*Cacatua goffini*)," *Proceedings of the Royal Society B: Biological Sciences* 281, no. 1793 (October 22, 2014): 1–8.

69. K. R. L. Hall and George B. Schaller, "Tool-using behavior of the California sea otter," *Journal of Mammalogy* 45, no. 2 (May 1964): 287–298.

70. Jessica A. Fujii, Katherine Ralls, and M. Tim Tinker, "Food abundance, prey morphology, and diet specialization influence individual sea otter tool use," *Behavioral Ecology* 28, no. 5 (September–October 2017): 1206–1216.

71. Rachel Smolker et al., "Sponge carrying by dolphins (*Delphinidae, Tursiops sp.*): A foraging specialization involving tool use?," *Ethology* 103, no. 6 (June 1997): 454–465.

72. Susan Milius, "Sponge moms: Dolphins learn tool use from their mothers," *Science News* 167 (June 11, 2005): 371–372.

73. Sonja Wild et al., "Integrating genetic, environmental, and social networks to reveal transmission pathways of a dolphin foraging innovation," *Current Biology* 30, no. 15 (August 2020): 3024–3030.

74. Vladimir Dinets, J. C. Brueggen, and J. D. Brueggen, "Crocodilians use tools for hunting," *Ethology Ecology & Evolution* 27, no. 1 (2015): 74–78.

75. Julian K. Finn, Tom Tregenza, and Mark D. Norman, "Defensive tool use in a coconut-carrying octopus," *Current Biology* 19, no. 23 (December 15, 2009): R1069–R1070.

76. David Scheel et al., "A second site occupied by *Octopus tetricus* at high densities, with notes on their ecology and behavior," *Marine and Freshwater Behaviour and Physiology* 50, no. 4 (September 2017): 285–291.

77. Crickette Sanz, Josep Call, and David Morgan, "Design complexity in termite-fishing tools of chimpanzees (*Pan troglodytes*)," *Biology Letters* 5, no. 3 (June 23, 2009): 293–296.

78. Mathias Osvath, "Spontaneous planning for future stone throwing by a male chimpanzee," *Current Biology* 19, no. 5 (March 10, 2009): R190–R191.

79. Hart et al., "Cognitive behaviour in Asian elephants," 839–847.

80. Lucas A. Bluff et al., "Tool use by wild New Caledonian crows *Corvus moneduloides* at natural foraging sites," *Proceedings of the Royal Society B: Biological Sciences* 277, no. 1686 (May 7, 2010): 1377–1385.

81. Anthony J. DeCasper et al., "Fetal reactions to recurrent maternal speech," *Infant Behavior and Development* 17, no. 2 (April–June 1994): 159–164.

82. Renaud Jardri et al., "Assessing fetal response to maternal speech using a noninvasive functional brain imaging technique," *International Journal of Developmental Neuroscience* 30, no. 2 (April 2012): 159–161.

83. Cecilia S. L. Lai et al., "A forkhead-domain gene is mutated in a severe speech and language disorder," *Nature* 413 (October 4, 2001): 519–523.

84. Michael Balter, "'Language gene' more active in young girls than boys," *Science* (February 2013).

85. W. Tecumseh Fitch, "The evolution of speech: A comparative review," *Trends in Cognitive Sciences* 4, no. 7 (July 1, 2000): 258–267.

86. D'Anastasio et al., "Micro-biomechanics of the Kebara 2 hyoid."

87. Johannes Krause et al., "The derived *FOXP2* variant of modern humans was shared with Neandertals," *Current Biology* 17, no. 21 (November 6, 2007): 1908–1912.

88. Adrien Meguerditchian, "Latéralité et communication gestuelle chez le babouin et le chimpanzé: À la recherche des précurseurs du langage" (PhD diss. on cognition, language, and education, University of Provence, Aix-Marseille 1, 2009).

89. Randall White, "Thoughts on social relationships and language in hominid evolution," *Journal of Social and Personal Relationships* 2, no. 1 (March 1985): 95–115.

90. Nicolas Mathevon, *Les animaux parlent: Sachons les écouter* (Paris: humenSciences, 2021).

91. Nicolas Mathevon et al., "What the hyena's laugh tells: Sex, age, dominance and individual signature in the giggling call of *Crocuta crocuta*," *BMC Ecology* 10, no. 9 (March 30, 2010).

92. Isabelle Charrier, Benjamin J. Pitcher, and Robert G. Harcourt, "Vocal recognition of mothers by Australian sea lion pups: Individual signature and environmental constraints," *Animal Behaviour* 78, no. 5 (November 2009): 1127–1134.

93. Pierre Jouventin, Thierry Aubin, and Thierry Lengagne, "Finding a parent in a king penguin colony: The acoustic system of individual recognition," *Animal Behaviour* 57, no. 6 (June 1999): 1175–1183.

94. Frédéric Sèbe et al., "Early vocal recognition of mother by lambs: Comparison of low and high-frequency vocalizations," *Animal Behaviour* 79, no. 5 (May 2010): 1055–1066.

95. Stephanie King and Vincent M. Janik, "Bottlenose dolphins can use learned vocal labels to address each other," *PNAS* 110, no. 32 (July 2013): 13216–13221.

96. D. Reby et al., "Individuality in the groans of fallow deer (*Dama dama*) bucks," *Journal of Zoology* 245, no. 1 (May 1998): 79–84.

97. Amélie L. Vergne et al., "Parent-offspring communication in the Nile crocodile *Crocodylus niloticus*: Do newborns' calls show an individual signature?," *Naturwissenschaften* 94, no. 1 (January 2007): 49–54.

98. Toshitaka N. Suzuki, "Semantic communication in birds: evidence from field research over the past two decades," *Ecological Research* 31, no. 3 (February 2016): 307–319.

99. Thomas T. Struhsaker, "Auditory communication among vervet monkeys (*Cercopithecus aethiops*)," in *Social Communication among Primates*, ed. Stuart A. Altmann (Chicago: University of Chicago Press, 1967), 281–324; Robert M. Seyfarth, Dorothy L. Cheney, and Peter Marler, "Vervet monkey alarm calls: Semantic communication in a free-ranging primate," *Animal Behaviour* 28, no. 4 (November 1980): 1070–1094.

100. Klaus Zuberbühler, "Referential labelling in Diana monkeys," *Animal Behaviour* 59, no. 5 (May 2000): 917–927.

101. Klaus Zuberbühler, "Predator-specific alarm calls in Campbell's monkeys, *Cercopithecus campbelli*," *Behavioral Ecology and Sociobiology* 50 (October 2001): 414–422.

102. C. N. Slobodchikoff et al., "Semantic information distinguishing individual predators in the alarm calls of Gunnison's prairie dogs," *Animal Behaviour* 42, no. 5 (November 1991): 713–719.

103. Zanna Clay and Klaus Zuberbühler, "Bonobos extract meaning from call sequences," *PLoS ONE* 6, no. 4 (April 27, 2011).

104. Catherine Crockford, Thibaud Gruber, and Klaus Zuberbühler, "Chimpanzee quiet hoo variants differ according to context," *Royal Society Open Science* 5, no. 5 (May 23, 2018): 172066.

105. Verena Maier, "Acoustic communication in the guinea fowl (*Numida meleagris*): Structure and use of vocalizations, and the principles of message coding," *Zeitschrift für Tierpsychologie* 59, no. 1 (January–December 1982): 29–83.

106. Christopher S. Evans, Linda Evans, and Peter Marler, "On the meaning of alarm calls: Functional reference in an avian vocal system," *Animal Behaviour* 46, no. 1 (July 1993): 23–38.

107. Verena Kersken et al., "A gestural repertoire of 1- to 2-year-old human children: In search of the ape gestures," *Animal Cognition* 22 (July 2019): 577–595.

108. Acoustical Society of America, "Environment and climate helped shape varied evolution of human languages," *ScienceDaily*, November 4, 2015.

109. C. N. Slobodchikoff and R. Coast, "Dialects in the alarm calls of prairie dogs," *Behavioral Ecology and Sociobiology* 7 (May 1980): 49–53.

110. Steven Green, "Dialects in Japanese monkeys: Vocal learning and cultural transmission of locale-specific vocal behavior?," *Zeitschrift für Tierpsychologie* 38, no. 3 (January–December 1975): 304–314.

111. Philippe Schlenker et al., "Monkey semantics: Two 'dialects' of Campbell's monkey alarm calls," *Linguistics and Philosophy* 37 (November 28, 2014): 439–501.

112. P. Marler and M. Tamura, "Song 'dialects' in three populations of white-crowned sparrows," *Condor* 64, no. 5 (September–October 1962): 368–377; Fernando Nottebohm, "The song of the chingolo, *Zonotrichia Capensis*, in

Argentina: Description and evaluation of a system of dialects," *Condor* 71, no. 3 (July 1, 1969): 299–315.

113. John K. B. Ford and H. D. Fisher, "Group-specific dialects of killer whales (*Orcinus orca*) in British Columbia," in *Communication and Behavior of Whales*, ed. Roger Payne (Boulder, CO: Westview Press, 1983), 129–161.

114. Burney J. Le Bœuf and Richard S. Peterson, "Dialects in elephant seals," *Science* 166, no. 3913 (December 26, 1969): 1654–1656.

115. Preston Somers, "Dialects in southern Rocky Mountain pikas, *Ochotona princeps* (Lagomorpha)," *Animal Behaviour* 21, no. 1 (February 1973): 124–137.

116. Vyacheslav A. Ryabov, "The study of acoustic signals and the supposed spoken language of the dolphins," *St. Petersburg Polytechnical University Journal: Physics and Mathematics* 2, no. 3 (October 2016): 231–239.

117. Jeremy Kuhn et al., "On the *-oo* suffix of Campbell's monkeys," *Linguistic Inquiry* 49, no. 1 (January 1, 2018): 169–181; Karim Ouattara, Alban Lemasson, and Klaus Zuberbühler, "Campbell's monkeys use affixation to alter call meaning," *PLoS ONE* 4, no. 11 (November 12, 2009): e7808.

118. Tanya Stivers et al., "Universals and cultural variation in turn-taking in conversation," *PNAS* 106, no. 26 (June 30, 2009): 10587–10592.

119. Bo Luo et al., "Brevity is prevalent in bat short-range communication," *Journal of Comparative Physiology A* 199, no. 4 (April 2013): 325–333.

120. Stuart Semple, Minna J. Hsu, and Govindasamy Agoramoorthy, "Efficiency of coding in macaque vocal communication," *Biology Letters* 6, no. 4 (August 23, 2010): 469–471.

121. Ramon Ferrer-i-Cancho and Antoni Hernández-Fernández, "The failure of the law of brevity in two New World primates: Statistical caveats," *Glottotheory* 4, no. 1 (April 2013): 45–55.

122. Morgan L. Gustison et al., "Gelada vocal sequences follow Menzerath's linguistic law," *PNAS* 113, no. 19 (May 10, 2016): E2750–E2758.

123. Raphaela Heesen et al., "Linguistic laws in chimpanzee gestural communication," *Proceedings of the Royal Society B: Biological Sciences* 286, no. 1896 (February 13, 2019): 20182900.

124. Duane M. Rumbaugh, ed., *Language Learning by a Chimpanzee: The Lana Project* (Cambridge: Academic Press, 2014).

125. Michael J. Beran et al., "A chimpanzee's (*Pan troglodytes*) long-term retention of lexigrams," *Animal Learning & Behavior* 28 (June 2000): 201–207.

126. Mark S. Seidenberg and Laura A. Petitto, "Communication, symbolic communication, and language: Comment on Savage-Rumbaugh, McDonald, Sevcik, Hopkins, and Rupert (1986)," *Journal of Experimental Psychology: General* 116, no. 3 (September 1987): 279–287.

127. E. S. Savage-Rumbaugh, "Communication, symbolic communication, and language: Reply to Seidenberg and Pettito," *Journal of Experimental Psychology: General* 116, no. 3 (September 1987): 288–292.

128. Adriano R. Lameira and Josep Call, "Time-space-displaced responses in the orangutan vocal system," *Science Advances* 4, no. 11 (November 14, 2018): eaau3401.

129. Gema Martin-Ordas et al., "Keeping track of time: Evidence for episodic-like memory in great apes," *Animal Cognition* 13, no. 2 (March 2010): 331–340.

130. Stephanie J. Babb and Jonathon D. Crystal, "Discrimination of what, when, and where is not based on time of day," *Learning & Behavior* 34 (May 2006): 124–130.

131. Amy-Lee Kouwenberg et al., "Episodic-like memory in crossbred Yucatan minipigs (*Sus scrofa*)," *Applied Animal Behaviour Science* 117, nos. 3–4 (March 2009): 165–172.

132. Claudia Fugazza, Ákos Pogány, and Ádám Miklósi, "Recall of others' actions after incidental encoding reveals episodic-like memory in dogs," *Current Biology* 26, no. 23 (December 5, 2016): 3209–3213.

133. Saho Takagi et al., "Use of incidentally encoded memory from a single experience in cats," *Behavioural Processes* 141, no. 3 (August 2017): 267–272.

134. Nicola S. Clayton and Anthony Dickinson, "Episodic-like memory during cache recovery by scrub jays," *Nature* 395 (September 17, 1998): 272–274.

135. Ann Zinkivskay, Farrah Nazir, and Tom V. Smulders, "What-where-when memory in magpies (*Pica pica*)," *Animal Cognition* 12, no. 1 (January 2009): 119–125.

136. Jonathan Henderson et al., "Timing in free-living rufous humming-birds, *Selasphorus rufus*," *Current Biology* 16, no. 5 (March 7, 2006): 512–515.

137. Christelle Jozet-Alves, Marion Bertin, and Nicola S. Clayton, "Evidence of episodic-like memory in cuttlefish," *Current Biology* 23, no. 23 (December 2, 2013): R1033–R1035.

138. Jennifer Hudin, "Did Alex have language?," *Ethics & Politics* 11, no. 1 (2009): 271–290.

139. H. Freyja Ólafsdóttir et al., "Hippocampal place cells construct reward related sequences through unexplored space," *eLife* 4 (June 26, 2015): e06063.

140. Daniel Baril, "La religion comme produit dérivé," *Religiologiques* 30 (Fall 2004): 141–158.

141. Sonya Kahlenberg and Richard Wrangham, "Toy Story," *BBC Wildlife Magazine*, January 2011, 52–57.

142. Sonya M. Kahlenberg and Richard W. Wrangham, "Sex differences in chimpanzees' use of sticks as play objects resemble those of children," *Current Biology* 20, no. 24 (December 21, 2010): R1067–R1068.

143. Hjalmar S. Kühl et al., "Chimpanzee accumulative stone throwing," *Scientific Reports* 6 (February 29, 2016): 1–8.

144. Carola Lentz and Hans-Jürgen Sturm, "Of trees and earth shrines: An interdisciplinary approach to settlement histories in the West African savanna," *History in Africa* 28 (January 2001): 139–168.

145. James B. Harrod, "The case for chimpanzee religion," *Journal for the Study of Religion, Nature and Culture* 8, no. 1 (2014): 8–45.

146. Jane Goodall, *In the Shadow of Man* (Boston: Houghton Mifflin Harcourt, 1971).

147. Sana Inoue and Tetsuro Matsuzawa, "Working memory of numerals in chimpanzees," *Current Biology* 17, no. 23 (December 4, 2007): R1004–R1005.

148. Francine L. Dolins et al., "Using virtual reality to investigate comparative spatial cognitive abilities in chimpanzees and humans," *American Journal of Primatology* 76, no. 5 (May 2014): 496–513.

149. Julia Watzek, Sarah M. Pope, and Sarah F. Brosnan, "Capuchin and rhesus monkeys but not humans show cognitive flexibility in an optional-switch task," *Scientific Reports* 9, no. 1 (September 13, 2019): 1–10.

150. Karri Neldner, Emma Collier-Baker, and Mark Nielsen, "Chimpanzees (*Pan troglodytes*) and human children (*Homo sapiens*) know when they are ignorant about the location of food," *Animal Cognition* 18, no. 3 (May 2015): 683–699.

151. J. D. Smith et al., "The uncertain response in the bottlenosed dolphin (*Tursiops truncatus*)," *Journal of Experimental Psychology: General* 124, no. 4 (December 1995): 391–408.

152. Gin Morgan et al., "Retrospective and prospective metacognitive judgments in rhesus macaques (*Macaca mulatta*)," *Animal Cognition* 17 (March 2014): 249–257.

153. Chikako Suda-King, "Do orangutans (*Pongo pygmaeus*) know when they do not remember?," *Animal Cognition* 11, no. 1 (January 2008): 21–42.

154. Allison L. Foote and Jonathon D. Crystal, "Metacognition in the rat," *Current Biology* 17, no. 6 (March 20, 2007): 551–555.

155. Wendy E. Shields et al., "Confidence judgments by humans and rhesus monkeys," *Journal of General Psychology* 132, no. 2 (April 2005): 165–186.

156. Kazuhiro Goto and Shigeru Watanabe, "Large-billed crows (*Corvus macrorhynchos*) have retrospective but not prospective metamemory," *Animal Cognition* 15, no. 1 (January 2012): 27–35.

157. Noriyuki Nakamura et al., "Do birds (pigeons and bantams) know how confident they are of their perceptual decisions?," *Animal Cognition* 14, no. 1 (January 2011): 83–93.

158. Brian Hare et al., "Chimpanzees know what conspecifics do and do not see," *Animal Behaviour* 59, no. 4 (April 2000): 771–785.

159. Thomas Bugnyar, Stephan A. Reber, and Cameron Buckner, "Ravens attribute visual access to unseen competitors," *Nature Communications* 7 (February 2, 2016): 1–6.

160. Joanna M. Dally, Nathan J. Emery, and Nicola S. Clayton, "Food-caching western scrub-jays keep track of who was watching when," *Science* 312, no. 5780 (June 16, 2006): 1662–1665.

161. Áam Miklósi et al., "A comparative study of the use of visual communicative signals in interactions between dogs (*Canis familiaris*) and humans and cats (*Felis catus*) and humans," *Journal of Comparative Psychology* 119, no. 2 (May 2005): 179–186.

162. Christopher Krupenye et al., "Great apes anticipate that other individuals will act according to false beliefs," *Science* 354, no. 6308 (October 7, 2016): 110–114.

163. Michael Tomasello et al., "Understanding and sharing intentions: The origins of cultural cognition," *Behavioral and Brain Sciences* 28, no. 5 (October 2005): 675–735.

164. Thomas Bugnyar and Bernd Heinrich, "Ravens, *Corvus corax*, differentiate between knowledgeable and ignorant competitors," *Proceedings of the Royal Society B: Biological Sciences* 272, no. 1573 (August 22, 2005): 1641–1646.

165. *Droit animal, éthique & sciences: Revue trimestrielle de la fondation LFDA* 94 (July 2017).

3. Our Bestial Emotions

166. Charles Darwin, *On the Origin of Species by Means of Natural Selection, or the Preservation of Favoured Races in the Struggle for Life* (London: John Murray, 1859).

167. Charles Darwin, *The Expression of the Emotions in Man and Animals* (London: John Murray, 1872).

168. Kristin Hagen and Donald M. Broom, "Emotional reactions to learning in cattle," *Applied Animal Behaviour Science* 85, nos. 3–4 (March 25, 2004): 203–213.

169. Gordon M. Burghardt, "The comparative reach of play and brain: Perspective, evidence, and implications," *American Journal of Play* 2 (Winter 2010): 338–356.

170. Annika S. Reinhold et al., "Behavioral and neural correlates of hide-and-seek in rats," *Science* 365, no. 6458 (September 13, 2019): 1180–1183.

171. Matthew C. O'Neill et al., "Chimpanzee super strength and human skeletal muscle evolution," *PNAS* 114, no. 28 (July 2017): 7343–7348.

172. Marc Levivier, "La fœtalisation de Louis Bolk," *Essaim* 26, no. 1 (June 2011): 153–168.

173. Jessica Serra, *Dans la tête d'un chat* (Paris: humenSciences, 2020).

174. Guillaume-Benjamin Duchenne, *Mécanisme de la physionomie humaine ou Analyse électro-physiologique de l'expression des passions* (Paris: J.-B. Baillière, 1876).

175. Yuu Mizuno, Hideko Takeshita, and Tetsuro Matsuzawa, "Behavior of infant chimpanzees during the night in the first 4 months of life: Smiling and suckling in relation to behavioral state," *Infancy* 9, no. 2 (March 2006): 221–240.

176. Fumito Kawakami, Masaki Tomonaga, and Juri Suzuki, "The first smile: Spontaneous smiles in newborn Japanese macaques (*Macaca fuscata*)," *Primates* 58, no. 1 (January 2017): 93–101.

177. Frans de Waal, *Mama's Last Hug: Animal Emotions and What They Tell Us about Ourselves* (New York: W. W. Norton, 2019).

178. Willibald Ruch and Paul Ekman, "The expressive pattern of laughter," in *Emotions, Qualia, and Consciousness*, ed. Alfred Kaszniak (Singapore: World Scientific Publishing, 2001): 426–443.

179. Moira Smith, "Laughter: Nature or culture?" (paper delivered at the 2008 meeting of the International Society for Humor Research, Alcala de Henares, Spain, 2008).

180. Miho Nagasawa et al., "Dogs can discriminate human smiling faces from blank expressions," *Animal Cognition* 14, no. 4 (July 2011): 525–533.

181. Bertrand L. Deputte and A. Doll, "Do dogs understand human facial expressions?," *Journal of Veterinary Behavior Clinical Applications and Research* 6, no. 1 (February 2011): 78–79.

182. Corsin A. Müller et al., "Dogs can discriminate emotional expressions of human faces," *Current Biology* 25, no. 5 (March 2, 2015): 601–605.

183. Steven Légaré, "Les origines évolutionnistes du rire et de l'humour" (PhD diss. in anthropology, University of Montreal, 2010).

184. Marina Davila Ross, Michael J. Owren, and Elke Zimmermann, "Reconstructing the evolution of laughter in great apes and humans," *Current Biology* 19, no. 13 (July 14, 2009): 1106–1111.

185. "Voilà ce qui se passe quand on chatouille un chimpanzé," *Gentside*, December 28, 2018, video, 0:36, https://www.maxisciences.com /chimpanze/voila-ce-qui-se-passe-quand-on-chatouille-un-chimpanze _art39157.html.

186. Robert R. Province, *Laughter: A Scientific Investigation* (New York: Viking, 2000).

187. J. Panksepp and J. Burgdorf, "Laughing rats? Playful tickling arouses high-frequency ultrasonic chirping in young rodents," *American Journal of Play* 2 (Winter 2010): 357–372.

188. S. Ishiyama and M. Brecht, "Neural correlates of ticklishness in the rat somatosensory cortex," *Science* 354, no. 6313 (November 11, 2016): 757–760.

189. Patricia Simonet, M. Murphy, and A. Lance, "Laughing dog: Vocalizations of domestic dogs during play encounters" (paper presented at the meeting of the Animal Behavior Society Corvallis, OR, 2001).

190. Patricia Simonet, Donna Versteeg, and Dan Storie, "Dog laughter: Recorded playback reduces stress related aggression in shelter dogs" (proceedings of the 7th International Conference on Environmental Enrichment, New York, 2005).

191. Gregory A. Bryant et al., "Detecting affiliation in colaughter across 24 societies," *PNAS* 113, no. 17 (April 26, 2016): 4682–4687.

192. V. Reddy, "Playing with others' expectations: Teasing and mucking about in the first year," in *Natural Theories of Mind: Evolution, Development and Simulation of Everyday Mindreading*, ed. Andrew Whiten (Oxford: Basil Blackwell, 1991).

193. Frans de Waal, *The Bonobo and the Atheist: In Search of Humanism among the Primates* (New York: W. W. Norton, 2013).

194. Jean de La Fontaine, "La jeune veuve," *Fables*, book VI, 1668.

195. Olivia Gazalé, *Le mythe de la virilité: Un piège pour les deux sexes* (Paris: Robert Laffont, 2017).

196. Homer, *Odyssey* (Paris: Hachette, 1999).

197. Anne Vincent-Buffault, *Histoire des larmes xviiie–xixe siècles* (Paris: Payot, 2001).

198. Juan Murube, "Basal, reflex, and psycho-emotional tears," *Ocular Surface* 7, no. 2 (April 2009): 60–66.

199. Ana Cláudia Raposo et al., "Comparative analysis of tear composition in humans, domestic mammals, reptiles, and birds," *Frontiers in Veterinary Science* 7 (May 22, 2020).

200. Rebecca E Doyle et al., "Administration of serotonin inhibitor p-Chlorophenylalanine induces pessimistic-like judgement bias in sheep," *Psychoneuroendocrinology* 36, no. 2 (February 2011): 279–288.

201. Jenny Stracke et al., "Serotonin depletion induces pessimistic-like behavior in a cognitive bias paradigm in pigs," *Physiology & Behavior* 174 (May 15, 2017): 18–26.

202. Rebecca E. Doyle et al., "Measuring judgement bias and emotional reactivity in sheep following long-term exposure to unpredictable and aversive events," *Physiology & Behavior* 102, no. 5 (March 28, 2011): 503–510.

203. Sandra Düpjan et al., "A design for studies on cognitive bias in the domestic pig," *Journal of Veterinary Behavior* 8, no. 6 (November–December 2013): 485–489.

204. J. Pittman and A. Piato, "Developing zebrafish depression-related models," in *The Rights and Wrongs of Zebrafish: Behavioral Phenotyping of Zebrafish*, ed. Allan V. Kalueff (Cham, Switzerland: Springer, 2017): 33–43.

205. Katherine A. Cronin et al., "Behavioral response of a chimpanzee mother toward her dead infant," *American Journal of Primatology* 73, no. 5 (May 2011): 415–421.

206. T. Matsuzawa, "The death of an infant chimpanzee at Bossou, Guinea," *Pan Africa News* 4 (June 1997): 4–6.

207. Edwin J. C. van Leeuwen, Katherine A. Cronin, and Daniel B. M. Haun, "Tool use for corpse cleaning in chimpanzees," *Scientific Reports* 7 (March 13, 2017): 44091.

208. William C. McGrew and Caroline E. G. Tutin, "Chimpanzee tool use in dental grooming," *Nature* 241 (February 16, 1973): 477–478.

209. F. Alves et al., "Supportive behavior of free-ranging Atlantic spotted dolphins (*Stenella frontalis*) toward dead neonates, with data

on perinatal mortality," *Acta Ethologica* 18 (November 2015): 301–304.

210. Iain Douglas-Hamilton et al., "Behavioural reactions of elephants towards a dying and deceased matriarch," *Applied Animal Behaviour Science* 100, no. 1–2 (October 2006): 87–102.

211. C. M. Bénard, "Esthétique," in *Dictionnaire des sciences philosophiques; Par une société de professeurs de philosophie*, ed. Adolphe Franck (Paris: Hachette, 1845): 479.

212. "Discours de Benoît XVI aux artistes (21 novembre)," Zenit.org: Le monde vu de Rome, November 23, 2009, https://fr.zenit.org/2009/11/23 /discours-de-benoit-xvi-aux-artistes-21-novembre/.

213. Stefano Ghirlanda, Liselotte Jansson, and Magnus Enquist, "Chickens prefer beautiful humans," *Human Nature* 13, no. 3 (September 2002): 383–389.

214. John A. Endler, "Bowerbirds, art and aesthetics: Are bowerbirds artists and do they have an aesthetic sense?," *Communicative & Integrative Biology* 5, no. 3 (May 1, 2012): 281–283.

215. John A. Endler, Lorna C. Endler, and Natalie R. Doerr, "Great bowerbirds create theaters with forced perspective when seen by their audience," *Current Biology* 20, no. 18 (September 28, 2010): 1679–1684; Laura A. Kelley and John A. Endler, "Illusions promote mating success in great bowerbirds," *Science* 335, no. 6066 (January 20, 2012): 335–338.

216. Roxana Torres and Alberto Velando, "A dynamic trait affects continuous pair assessment in the blue-footed booby, *Sula nebouxii*," *Behavioral Ecology and Sociobiology* 55 (August 9, 2003): 65–72.

217. Eric Vallet and Michel Kreutzer, "Female canaries are sexually responsive to special song phrases," *Animal Behaviour* 49, no. 6 (June 1995): 1603–1610.

218. K. Payne, "The progressively changing songs of humpback whales: A window on the creative process in a wild animal," in *The Origins of Music*, eds. Nils L. Wallin, Björn Merker, and Steven Brown (Cambridge, MA: MIT Press, 2000), 135–150.

219. Michel Kreutzer, "Du choix esthétique chez les animaux," *Revue d'esthétique* 40 (2001): 113–116.

220. Christopher S. Henshilwood et al., "An abstract drawing from the 73,000-year-old levels at Blombos Cave, South Africa," *Nature* 562, no. 7725 (September 2018): 115–118.

221. Anne-Marie Christin, "Les origines de l'écriture: Image, signe, trace," *Le Débat* 106 (September–October 1999): 28–36.

222. Charles Baudelaire, "La musique," *Les fleurs du mal* (Paris: Flammarion, 2011).

223. Aniruddh D. Patel et al., "Experimental evidence for synchronization to a musical beat in a nonhuman animal," *Current Biology* 19, no. 10 (May 26, 2009): 827–830.

224. "Spontaneity and diversity of movement to music are not uniquely human," *Current Biology*, July 8, 2019, videos, 1:18, 1:20, https://www.sciencedirect.com/science/article/pii/S0960982219306049.

225. R. Joanne Jao Keehn et al., "Spontaneity and diversity of movement to music are not uniquely human," *Current Biology* 29, no. 13 (July 8, 2019): R621–R622.

226. A. Bowman et al., "'Four Seasons' in an animal rescue centre: Classical music reduces environmental stress in kennelled dogs," *Physiology & Behavior* 143 (May 1, 2015): 70–82.

227. J. L. Campo, M. G. Gil, and S. G. Dávila, "Effects of specific noise and music stimuli on stress and fear levels of laying hens of several breeds," *Applied Animal Behaviour Science* 91, no. 1–2 (May 2005): 75–84.

228. S. E. Papoutsoglou et al., "Effect of Mozart's music (Romanze-Andante of 'Eine Kleine Nacht Musik,' sol major, K525) stimulus on common carp (*Cyprinus carpio* L.) physiology under different light conditions," *Aquacultural Engineering* 36, no. 1 (January 2007): 61–72.

229. Frances Rauscher, Desix Robinson, and Jason Jens, "Improved maze learning through early music exposure in rats," *Neurological Research* 20, no. 5 (March 1998): 427–432.

230. A. J. Blood and R. J. Zatorre, "Intensely pleasurable responses to music correlate with activity in brain regions implicated in reward and emotion," *PNAS* 98, no. 20 (September 25, 2001): 11818–11823.

231. Emmanuel Bigand and Barbara Tillmann, *La symphonie neuronale: Pourquoi la musique est indispensable au cerveau* (Paris: humenSciences, 2020).

232. P. G. Hepper, D. Scott, and S. Shahidullah, "Newborn and fetal response to maternal voice," *Journal of Reproductive and Infant Psychology* 11, no. 3 (1993): 147–153.

233. Berna Alay and Figen Isik Esenay, "The clinical effect of classical music and lullaby on term babies in neonatal intensive care unit: A randomised controlled trial," *Journal of the Pakistan Medical Association* 69, no. 4 (April 2019): 459–463.

234. Giuseppina Persico et al., "Maternal singing of lullabies during pregnancy and after birth: Effects on mother-infant bonding and on newborns' behaviour. Concurrent Cohort Study," *Women and Birth* 30, no. 4 (August 2017): e214–e220.

235. Anne Fernald and Thomas Simon, "Expanded intonation contours in mothers' speech to newborns," *Developmental Psychology* 20, no. 1 (January 1984): 104–113.

236. Birgit Mampe et al., "Newborns' cry melody is shaped by their native language," *Current Biology* 19, no. 23 (December 15, 2009): 1994–1997.

237. Isabelle Peretz, *Apprendre la musique: Nouvelles des neurosciences* (Paris: Odile Jacob, 2018).

4. Vices and Virtues

238. Julie Lacaze, "Pourquoi des enfants et des lamas étaient-ils sacrifiés par le peuple Chimú?," *National Geographic*, February 2019.

239. Alexandre Martin, "Historique des droits de l'homme," *Le Monde*, August 14, 2003.

240. Azim F. Shariff, Jared Piazza, and Stephanie R. Kramer, "Morality and the religious mind: Why theists and nontheists differ," *Trends in Cognitive Sciences* 18, no. 9 (September 2014): 439–441.

241. Christine Tamir, Aidan Connaughton, and Ariana Monique Salazar, "The Global God Divide," Global Attitudes & Trends (Washington, DC: Pew Research Center, July 20, 2020).

242. Ronit Roth-Hanania, Maayan Davidov, and Carolyn Zahn-Waxler, "Empathy development from 8 to 16 months: Early signs of concern for others," *Infant Behavior and Development* 34, no. 3 (June 2011): 447–458.

243. Mitzi-Jane E. Liddle, Ben S. Bradley, and Andrew Mcgrath, "Baby empathy: Infant distress and peer prosocial responses," *Infant Mental Health Journal* 36, no. 4 (June 2015): 446–458.

244. The results quoted are those from a self-completed questionnaire. Varun Warrier et al., "Genome-wide analyses of self-reported empathy: Correlations with autism, schizophrenia, and anorexia nervosa," *Translational Psychiatry* 8 (December 2018): 1–10.

245. Simon Baron-Cohen and Sally Wheelwright, "The Empathy Quotient: An investigation of adults with Asperger syndrome or high functioning autism, and normal sex differences," *Journal of Autism and Developmental Disorders* 34, no. 2 (April 2004): 163–175.

246. Birgit Derntl et al., "How specific are emotional deficits? A comparison of empathic abilities in schizophrenia, bipolar and depressed patients," *Schizophrenia Research* 142, no. 1–3 (December 2012): 58–64.

247. Tania M. Michaels et al., "Cognitive empathy contributes to poor social functioning in schizophrenia: Evidence from a new self-report measure of cognitive and affective empathy," *Psychiatry Research* 220, no. 3 (December 30, 2014): 803–810.

248. Michael James Weightman, Tracy Michele Air, and Bernhard Theodor Baune, "A review of the role of social cognition in major depressive disorder," *Frontiers in Psychiatry* 5 (December 11, 2014): 179.

249. Erik Trinkaus and Sébastien Villotte, "External auditory exostoses and hearing loss in the Shanidar 1 Neandertal," *PLoS ONE* 12, no. 10 (October 20, 2017): e0186684.

250. R. M. Church, "Emotional reactions of rats to the pain of others," *Journal of Comparative and Physiological Psychology* 52, no. 2 (1959): 132–134.

251. Ewelina Knapska et al., "Social modulation of learning in rats," *Learning & Memory* 17, no. 1 (December 2010): 35–42; Piray Atsak et al., "Experience modulates vicarious freezing in rats: A model for empathy," *PLoS ONE* 6, no. 7 (July 2011): e21855; Inbal Ben-Ami Bartal, Jean Decety, and Peggy Mason, "Empathy and pro-social behavior in rats," *Science* 334, no. 6061 (December 9, 2011): 1427–1430; Chun-Li Li et al., "Validating rat model of empathy for pain: Effects of pain expressions in social partners," *Frontiers in Behavioral Neuroscience* 12, no. 242 (October 17, 2018): 242.

252. Cristina Gonzalez-Liencres, Simone G. Shamay-Tsoory, and Martin Brüne, "Towards a neuroscience of empathy: Ontogeny, phylogeny, brain mechanisms, context and psychopathology," *Neuroscience & Biobehavioral Reviews* 37, no. 8 (September 2013): 1537–1548.

253. J. L. Edgar et al., "Avian maternal response to chick distress," *Proceedings of the Royal Society B: Biological Sciences* 278, no. 1721 (October 22, 2011): 3129–3134.

254. Peter Kropotkin, *Mutual Aid: A Factor of Evolution* (New York: McClure Phillips & Co., 1902).

255. Charles Darwin, *The Descent of Man, and Selection in Relation to Sex* (London: John Murray, 1871).

256. Edward O. Wilson, *Sociobiology: The New Synthesis* (Cambridge, MA: Harvard University Press, 2000).

257. George E. Rice and Priscilla Gainer, "'Altruism' in the albino rat," *Journal of Comparative and Physiological Psychology* 55, no. 1 (1962): 123–125.

258. Venkat R. Lakshminarayanan and Laurie R. Santos, "Capuchin monkeys are sensitive to others' welfare," *Current Biology* 18, no. 21 (November 11, 2008): R999–R1000.

259. Douglas-Hamilton et al., "Behavioural reactions of elephants towards a dying and deceased matriarch," 87–102.

260. Lucy A. Bates et al., "Do elephants show empathy?," *Journal of Consciousness Studies* 15, nos. 10–11 (January 2008): 204–225.

261. Deborah Custance and Jennifer Mayer, "Empathic-like responding by domestic dogs (*Canis familiaris*) to distress in humans: An exploratory study," *Animal Cognition* 15, no. 5 (September 2012): 851–859.

262. Elisabetta Palagi, Tommaso Paoli, and Silvana Borgognini Tarli, "Reconciliation and consolation in captive bonobos (*Pan paniscus*)," *American Journal of Primatology* 62, no. 1 (January 2004): 15–30.

263. Frans B. M. de Waal and Angeline van Roosmalen, "Reconciliation and consolation among chimpanzees," *Behavioral Ecology and Sociobiology* 5 (March 1979): 55–66; Roman M. Wittig and Christophe Boesch, "The choice of post-conflict interactions in wild chimpanzees (*Pan troglodytes*)," *Behaviour* 140, no. 11 (November-December 2003): 1527–1559.

264. Giada Cordoni, Elisabetta Palagi, and Silvana Borgognini Tarli, "Reconciliation and consolation in captive western gorillas," *International Journal of Primatology* 27 (November 2006): 1365–1382; S. Mallavarapu et al., "Postconflict behavior in captive western lowland gorillas (*Gorilla gorilla gorilla*)," *American Journal of Primatology* 68, no. 8 (August 2006): 789–801; Frans de Waal, *De la réconciliation chez les primates* (*Peacemaking among Primates*) (Paris: Flammarion, 1992).

265. Joshua M. Plotnik and Frans B. M. de Waal, "Asian elephants (*Elephas maximus*) reassure others in distress," *PeerJ* 2 (February 18, 2014).

266. Amanda M. Seed, Nicola S. Clayton, and Nathan J. Emery, "Postconflict third-party affiliation in rooks, *Corvus frugilegus*," *Current Biology* 17, no. 2 (January 23, 2007): 152–158.

267. Elisabetta Palagi and Giada Cordoni, "Postconflict third-party affiliation in *Canis lupus*: Do wolves share similarities with the great apes?," *Animal Behaviour* 78, no. 4 (October 2009): 979–986.

268. Annemieke K. A. Cools, Alain J.-M. Van Hout, and Mark H. J. Nelissen, "Canine reconciliation and third-party-initiated post-conflict affiliation: Do peacemaking social mechanisms in dogs rival those of higher primates?," *Ethology* 114, no. 1 (January 2008): 53–63; Mylene Quervel-Chaumette et al., "Investigating empathy-like responding to conspecifics' distress in pet dogs," *PLoS ONE* 11, no. 4 (April 28, 2016): e0152920.

269. J. P. Burkett et al., "Oxytocin-dependent consolation behavior in rodents," *Science* 351, no. 6271 (January 22, 2016): 375–378.

270. Georges Chapouthier, "L'homme, un pont entre deux mondes: Nature et culture," *Le Philosophoire* 2, no. 23 (January 2004): 99–114.

271. H. Rae Westbury and David L. Neumann, "Empathy-related responses to moving film stimuli depicting human and non-human animal targets in negative circumstances," *Biological Psychology* 78, no. 1 (April 2008): 66–74.

272. Alexandre Baratta, "La face sombre de l'empathie," *Atlantico*, September 29, 2015.

273. Margaret C. Crofoot and Richard W. Wrangham, "Intergroup Aggression in Primates and Humans: The Case for a Unified Theory," in *Mind the Gap:*

Tracing the Origins of Human Universals, eds. Peter M. Kappeler and Joan Silk (Berlin: Springer, 2009): 171–197.

274. Boban Docevski, "The Gombe War of Tanzania: A four-year-long guerrilla war between two groups of chimpanzees," *Vintage News*, August 31, 2017.

275. Jane Goodall, *Through a Window: My Thirty Years with the Chimpanzees of Gombe* (Boston: Houghton Mifflin Harcourt, 2010).

276. David P. Watts et al., "Lethal intergroup aggression by chimpanzees in Kibale National Park, Uganda," *American Journal of Primatology* 68, no. 2 (February 2006): 161–180.

277. R. Brian Ferguson, "Born to Live: Challenging Killer Myths," in *Origins of Altruism and Cooperation*, eds. Robert W. Sussman and C. Robert Cloninger (New York: Springer, 2011): 249–270.

278. Hogan M. Sherrow and Sylvia J. Amsler, "New intercommunity infanticides by the chimpanzees of Ngogo, Kibale National Park, Uganda," *International Journal of Primatology* 28 (March 13, 2007): 9–22.

279. Richard W. Wrangham and Luke Glowacki, "Intergroup aggression in chimpanzees and war in nomadic hunter-gatherers: Evaluating the chimpanzee model," *Human Nature* 23 (March 3, 2012): 5–29.

280. Jill D. Pruetz et al., "Intragroup lethal aggression in West African chimpanzees (*Pan troglodytes verus*): Inferred killing of a former alpha male at Fongoli, Senegal," *International Journal of Primatology* 38 (January 23, 2017): 31–57.

281. M. Wilson et al., "Rates of lethal aggression in chimpanzees depend on the number of adult males rather than measures of human disturbance," *Nature* 513 (September 2014): 414–417.

282. Maurice Maeterlinck, *La vie des fourmis* (Archipoche, 2020). Bernard Werber is the author of a trilogy of novels on ants (Paris, Albin Michel): *Les fourmis*, 1991; *Le jour des fourmis*, 1992; *La révolution des fourmis*, 1996.

283. Aiming Zhou, Yuzhe Du, and Jian Chen, "Ants adjust their tool use strategy in response to foraging risk," *Functional Ecology* (October 2020): 1–12.

284. Alexandra Achenbach and Susanne Foitzik, "First evidence for slave rebellion: Enslaved ant workers systematically kill the brood of their

social parasite *Protomognathus americanus*," *Evolution: International Journal of Organic Evolution* 63, no. 4 (April 2009): 1068–1075.

285. Alice Laciny et al., "*Colobopsis explodens* sp. n., model species for studies on 'exploding ants' (Hymenoptera, Formicidae), with biological notes and first illustrations of males of the *Colobopsis cylindrica* group," *ZooKeys* 751 (April 19, 2018): 1–40.

286. Erik T. Frank, M. Wehrhahn, and K. Linsenmair, "Wound treatment and selective help in a termite-hunting ant," *Proceedings of the Royal Society B: Biological Sciences* 285, no. 1872 (February 14, 2018): 20172457.

287. Marc-Antoine Pelaez, "Djebel Sahaba, lieu du premier massacre humain connu à ce jour," May 13, 2016, https://marcantoinepelaez.wordpress .com/2016/05/13/djebel-sahaba-lieu-du-premier-massacre-humain -connu-a-ce-jour/.

288. M. Mirazón Lahr et al., "Inter-group violence among early Holocene hunter-gatherers of West Turkana, Kenya," *Nature* 529 (January 20, 2016): 394–398.

289. Donatella Usai, "The Qadan, the Jebel Sahaba Cemetery and the Lithic Collection," *Archaeologia Polona* 58 (July 2020): 99–119.

290. Kent V. Flannery and Joyce Marcus, "A New World Perspective on the 'Death' of Archaeological Theory," in *The Death of Archaeological Theory?*, eds. John Bintliff and Mark Pearce (Oxford: Oxbow, 2011), 23–30.

291. R. Brian Ferguson, "Violence and War in Prehistory," in *Troubled Times: Violence and Warfare in the Past*, eds. David W. Frayer and Debra L. Martin (Amsterdam: Psychology Press, 1997).

292. Konrad Lorenz, *L'agression: Une histoire naturelle du mal* (Paris: Flammarion, 1993); *L'envers du miroir: Une histoire naturelle de la connaissance* (Paris: Flammarion, 2010); *Les fondements de l'éthologie* (Paris: Flammarion, 2009).

293. Raymond A. Dart, "The predatory transition from ape to man," *International Anthropological and Linguistic Review* 1, no. 4 (1953): 201–218.

294. C. K. Brain, *The Hunters or the Hunted? An Introduction to African Cave Taphonomy* (Chicago: University of Chicago Press, 1981); Matt Cartmill, *A View to a Death in the Morning: Hunting and Nature through History* (Cambridge, MA: Harvard University Press, 1996).

295. R. W. Wrangham, "Evolution of coalitionary killing," *American Journal of Physical Anthropology* 110 (December 1999): 1–30.

296. I. J. N. Thorpe, "The Ancient Origins of Warfare and Violence," in *Warfare, Violence and Slavery in Prehistory*, ed. Mike Parker Pearson (Oxford: British Archaeological Reports Publishing, 2005), 1–18.

297. Frans de Waal, *Peacemaking among Primates* (Cambridge, MA: Harvard University Press, 1989).

298. Robert H. Waterson, Eric S. Lander, and Richard K. Wilson, "Initial sequence of the chimpanzee genome and comparison with the human genome," *Nature* 437 (September 1, 2005): 69–87.

299. Kay Prüfer et al., "The bonobo genome compared with the chimpanzee and human genomes," *Nature* 486 (June 13, 2012): 527–531.

300. Ryuichi Sakate et al., "Mapping of chimpanzee full-length cDNAs onto the human genome unveils large potential divergence of the transcriptome," *Gene* 399, no. 1 (September 1, 2007): 1–10.

301. Robert Knox Dentan, *The Semai: A Nonviolent People of Malaya* (New York: Holt, Rinehart and Winston, 1997).

302. Clayton A. Robarchek and Carole J. Robarchek, "Reciprocities and realities: World views, peacefulness, and violence among Semai and Waorani," *Aggressive Behavior* 24, no. 2 (December 1998): 123–133.

303. Geoffrey Moss, "Explaining the absence of violent crime among the Semai of Malaysia: Is criminological theory up to the task?," *Journal of Criminal Justice* 25, no. 3 (1997): 177–194.

304. Paul L. Koch and Anthony D. Barnosky, "Late Quaternary extinctions: State of the debate," *Annual Review of Ecology, Evolution, and Systematics* 37 (December 12, 2006): 215–250.

5. Sex Machine?

305. Stephanie Ortigue et al., "Neuroimaging of love: fMRI meta-analysis evidence toward new perspectives in sexual medicine," *Journal of Sexual Medicine* 7, no. 11 (November 2010): 3541–3552.

306. John P. McGann, "Poor human olfaction is a 19th-century myth," *Science* 356, no. 6338 (May 12, 2017): eaam7263.

307. C. Wedekind et al., "MHC-dependent mate preferences in humans," *Proceedings of the Royal Society B: Biological Sciences* 260, no. 1359 (June 22, 1995): 245–249.

308. Jessica M. Gaby and Pamela Dalton, "Discrimination between individual body odors is unaffected by perfume," *Perception* 48, no. 11 (November 2019): 1104–1123.

309. J. M. Setchell et al., "Opposites attract: MHC-associated mate choice in a polygynous primate," *Journal of Evolutionary Biology* 23, no. 1 (January 2010): 136–148.

310. K. Yamazaki et al., "Control of mating preferences in mice by genes in the major histocompatibility complex," *Journal of Experimental Medicine* 144, no. 5 (November 2, 1976): 1324–1335; Wayne K. Potts, C. Jo Manning, and Edward K. Wakeland, "Mating patterns in seminatural populations of mice influenced by MHC genotype," *Nature* 352 (August 1991): 619–621.

311. C. Landry et al., "'Good genes as heterozygosity': The major histocompatibility complex and mate choice in Atlantic salmon (*Salmo salar*)," *Proceedings of the Royal Society B: Biological Sciences* 268, no. 1473 (June 22, 2001): 1279–1285; Muna Agbali et al., "Mate choice for nonadditive genetic benefits correlate with MHC dissimilarity in the rose bitterling (*Rhodeus ocellatus*)," *Evolution* 64, no. 6 (June 2010): 1683–1696.

312. Mats Olsson et al., "Major histocompatibility complex and mate choice in sand lizards," *Proceedings of the Royal Society B: Biological Sciences* 270, Supplement 2 (November 7, 2003): S254–S256.

313. Matteo Griggio et al., "Female house sparrows 'count on' male genes: Experimental evidence for MHC-dependent mate preference in birds," *BMC Evolutionary Biology* 11 (February 14, 2011): 44.

314. Marylène Boulet, Marie J. E. Charpentier, and Christine M. Drea, "Decoding an olfactory mechanism of kin recognition and inbreeding avoidance in a primate," *BMC Evolutionary Biology* 9 (December 3, 2009): 281.

315. Palestina Guevara-Fiore, Jessica Stapley, and Penelope J. Watt, "Mating effort and female receptivity: How do male guppies decide when to

invest in sex?," *Behavioral Ecology and Sociobiology* 64 (May 2010): 1665–1672.

316. Robert T. Mason and M. Rockwell Parker, "Social behavior and phero-monal communication in reptiles," *Journal of Comparative Physiology A* 196 (June 29, 2010): 729–749.

317. R. T. Mason et al., "Characterization, synthesis, and behavioral responses to the sex attractiveness pheromones of the red-sided garter snake (*Thamnophis sirtalis parietalis*)," *Journal of Chemical Ecology* 16, no. 7 (July 1990): 2353–2369.

318. Samuel P. Caro and Jacques Balthazart, "Pheromones in birds: Myth or reality?," *Journal of Comparative Physiology A* 196, no. 10 (October 2010): 751–766.

319. Sachiko Haga et al., "The male mouse pheromone ESP1 enhances female sexual receptive behavior through a specific vomeronasal receptor," *Nature* 466 (July 1, 2010): 118–122.

320. Andreas Keller et al., "Genetic variation in a human odorant receptor alters odour perception," *Nature* 449 (September 16, 2007): 468–472.

321. S. L. Black and C. Biron, "Androstenol as a human pheromone: No effect on perceived physical attractiveness," *Behavioral and Neural Biology* 34, no. 3 (March 1982): 326–330; A. R. Gustavson, M. E. Dawson, and D. G. Bonett, "Androstenol, a putative human pheromone, affects human (*Homo sapiens*) male choice performance," *Journal of Comparative Psychology* 101, no. 2 (June 1987): 210–212.

322. Roslyn Dakin et al., "Biomechanics of the peacock's display: How feather structure and resonance influence multimodal signaling," *PLoS ONE* 11, no. 4 (April 27, 2016).

323. Dalila Bovet, "Un paon très vantard . . . ," *Cerveau & Psycho* 56 (March 7, 2013): 92–93.

324. A. T. Bennett and I. C. Cuthill, "Ultraviolet vision in birds: What is its function?," *Vision Research* 34, no. 11 (June 1994): 1471–1478.

325. "*The Mating Mind: How Sexual Choice Shaped the Evolution of Human Nature* by Geoffrey Miller, New York: Doubleday, 2000. Reviewed by John D. Wagner," *Human Nature Review* 2 (March 2002): 110–113.

326. Viren Swami and Martin J. Tovée, "Resource security impacts men's female breast size preferences," *PLoS ONE* 8, no. 3 (March 6, 2013): e57623.

327. Desmond Morris, *The Naked Ape* (London: Jonathan Cape, 1967).

328. Sylvie Bailly, *Des siècles de beauté: Entre séduction et politique* (Brussels: Primento, 2015), 121.

329. Pauline Boulet, "Les hommes et le maquillage: Une démocratisation difficile," *L'Express*, July 26, 2019.

330. David A. Puts, Benedict C. Jones, and Lisa M. DeBruine, "Sexual selection on human faces and voices," *Journal of Sex Research* 49, no. 2–3 (March 2012): 227–243.

331. Steven W. Gangestad and Randy Thornhill, "Human oestrus," *Proceedings of the Royal Society B: Biological Sciences* 275, no. 1638 (May 7, 2008): 991–1000; Benedict C. Jones et al., "Effects of menstrual cycle phase on face preferences," *Archives of Sexual Behavior* 37, no. 1 (January 2008): 78–84.

332. Charles Darwin, *The Expression of the Emotions in Man and Animals* (London: John Murray, 1872).

333. H. E. Ross et al., "Characterization of the oxytocin system regulating affiliative behavior in female prairie voles," *Neuroscience* 162, no. 4 (September 15, 2009): 892–903.

334. Chloé Laubu, Philippe Louâpre, and François-Xavier Dechaume-Moncharmont, "Pair-bonding influences affective state in a monogamous fish species," *Proceedings of the Royal Society B: Biological Sciences* 286, no. 1904 (June 12, 2019).

335. Frans de Waal, *Are We Smart Enough to Know How Smart Animals Are?* (New York: W. W. Norton & Company, 2016).

336. Sheril Kirshenbaum, *The Science of Kissing: What Our Lips Are Telling Us* (New York: Grand Central Publishing, 2011).

337. Remco Kort et al., "Shaping the oral microbiota through intimate kissing," *Microbiome* 2 (November 17, 2014): 41.

338. Rafael Wlodarski and Robin I. M. Dunbar, "Examining the possible functions of kissing in romantic relationships," *Archives of Sexual Behavior* 42, no. 8 (November 2013): 1415–1423.

339. Sandra Murphy and Polly Dalton, "Out of touch? Visual load induces inattentional numbness," *Journal of Experimental Psychology: Human Perception and Performance* 42, no. 6 (June 2016): 761–765.

340. Krista Marie McLennan, "Social bonds in dairy cattle: The effect of dynamic group systems on welfare and productivity" (PhD diss., University of Northampton, 2013).

341. Blake Edgar, "La monogamie, un atout pour notre espèce," *Pour la science* 445 (October 23, 2014).

342. Christopher Opie et al., "Male infanticide leads to social monogamy in primates," *PNAS* 110, no. 33 (August 2013): 13328–13332.

343. Devra G. Kleiman, "Monogamy in mammals," *Quarterly Review of Biology* 52, no. 1 (March 1977): 39–69.

344. Peter N. M. Brotherton and Petr E. Komers, "Mate guarding and the evolution of social monogamy in mammals," in *Monogamy: Mating Strategies and Partnerships in Birds, Humans and Other Mammals*, eds. Ulrich H. Reichard and Christophe Boesch (Cambridge: Cambridge University Press, 2003): 42–58; P. E. Komers and P. N. Brotherton, "Female space use is the best predictor of monogamy in mammals," *Proceedings of the Royal Society B: Biological Sciences* 264, no. 1386 (September 22, 1997): 1261–1270.

345. C. P. Van Schaik and R. I. M. Dunbar, "The evolution of monogamy in large primates: A new hypothesis and some crucial tests," *Behaviour* 115, no. 1–2 (January 1990): 30–61.

346. J. D. Bygott, "Cannibalism among Wild Chimpanzees," *Nature* 238 (August 18, 1972): 410–411; J. Goodall, "Infant killing and cannibalism in free-living chimpanzees," *Folia Primatologica* 28, no. 4 (1977): 259–282; Toshisada Nishida and Kenji Kawanaka, "Within-group cannibalism by adult male chimpanzees," *Primates* 26 (July 1985): 274–284; Yukio Takahata, "Adult male chimpanzees kill and eat a male newborn infant: Newly observed intragroup infanticide and cannibalism in Mahale National Park, Tanzania," *Folia Primatologica* 44, no. 3–4 (January 1985): 161–170; David P. Watts and John C. Mitani, "Infanticide and cannibalism by male chimpanzees at Ngogo, Kibale National Park, Uganda," *Primates* 41, no. 4 (October 2000): 357–365.

347. Adriana E. Lowe, Catherine Hobaiter, and Nicholas E. Newton-Fisher, "Countering infanticide: Chimpanzee mothers are sensitive to the relative risks posed by males on differing rank trajectories," *American Journal of Physical Anthropology* 168, no. 1 (January 2019): 3–9.

348. L. J. Pitkow et al., "Facilitation of affiliation and pair-bond formation by vasopressin receptor gene transfer into the ventral forebrain of a monogamous vole," *Journal of Neuroscience* 21, no. 18 (September 15, 2001): 7392–7396.

349. Hasse Walum et al., "Genetic variation in the vasopressin receptor 1a gene (*AVPR1A*) associates with pair-bonding behavior in humans," *PNAS* 105, no. 37 (September 16, 2008): 14153–14156.

350. Hasse Walum et al., "Variation in the oxytocin receptor gene is associated with pair-bonding and social behavior," *Biological Psychiatry* 71, no. 5 (March 1, 2012): 419–426.

351. Dirk Scheele et al., "Oxytocin modulates social distance between males and females," *Journal of Neuroscience* 32, no. 46 (November 14, 2012): 16074–16079.

352. Michael J. Lyons et al., "A twin study of sexual behavior in men," *Archives of Sexual Behavior* 33, no. 2 (April 2004): 129–136.

353. Serge Wunsch, "L'influence de la cognition sur la sexualité," *Sexologies* 26, no. 1 (January–March 2017): 36–43.

354. Serge Wunsch, *Comprendre les origines de la sexualité humaine: Neurosciences, éthologie, anthropologie* (Paris: L'Esprit du temps, 2014).

355. Bruce Bagemihl, *Biological Exuberance: Animal Homosexuality and Natural Diversity* (London: Profile Books, 1999).

356. Janlou Chaput, "Chez les animaux aussi, quand il y a du sexe, il y a du plaisir!" *Futura Sciences*, February 26, 2013.

357. Sigmund Freud, *Trois essais sur la théorie sexuelle: 1905–1924* (Paris: Flammarion, 2011).

358. Élodie Lavigne, "Après une excision, le plaisir sexuel reste possible," *Le Matin*, February 26, 2017.

359. United Nations Children's Fund (UNICEF), "Female Genital Mutilation/Cutting: A statistical overview and exploration of the dynamics of

change," July 2013, https://data.unicef.org/resources/fgm-statistical
-overview-and-dynamics-of-change/.

360. Lavigne, "Après une excision, le plaisir sexuel reste possible."

361. Dara N. Orbach and Patricia L. R. Brennan, "Functional morphology
of the dolphin clitoris," *FASEB Journal* 33, no. S1 (April 2019): 10–14.

362. D. A. Goldfoot et al., "Behavioral and physiological evidence of sexual
climax in the female stump-tailed macaque (*Macaca arctoides*)," *Science*
208, no. 4451 (June 27, 1980): 1477–1479.

363. A. Troisi A and M. Carosi, "Female orgasm rate increases with male
dominance in Japanese macaques," *Animal Behaviour* 56, no. 5
(November 1998): 1261–1266.

364. Frans B. M. de Waal, "Bonobo sex and society: The behavior of a close
relative challenges assumptions about male supremacy in human
evolution," *Scientific American* 272 (March 1995): 82–88; Palagi, Paoli, and
Tarli, "Reconciliation and consolation in captive bonobos (*Pan paniscus*),"
15–30.

365. Janlou Chaput, "Sexualité: Les animaux ne manquent pas de pratiques!,"
Futura Sciences, February 4, 2013.

366. Min Tan et al., "Fellatio by fruit bats prolongs copulation time," *PLoS ONE*
4, no. 10 (October 2009).

367. Bagemihl, *Biological Exuberance*.

368. "Catechism of the Catholic Church," official Vatican website, https://www
.vatican.va/archive/ENG0015/_INDEX.HTM.

369. Pierre Humbert and Jérôme Palazzolo, *Petite histoire de la masturbation*
(Paris: Odile Jacob, 2009).

370. R. Thomsen and V. Sommer, "Masturbation (nonhuman primates)," in *The
International Encyclopedia of Human Sexuality*, ed. Patricia Whelehan and
Anne Bolin (Hoboken, NJ: John Wiley & Sons, 2015), 721–817.

371. Frans de Waal and Frans Lanting, *Bonobo: The Forgotten Ape* (Berkeley:
University of California Press, 1997).

372. Ruth Thomsen and Joseph Soltis, "Male masturbation in free-ranging
Japanese macaques," *International Journal of Primatology* 25 (October
2004): 1033–1041.

373. Morgane Kergoat, "VIDEO: Un dauphin harcèle sexuellement une plongeuse," *Sciences et Avenir*, March 16, 2015.

374. Arnaud Truchet, "Homosexualité: le pape François favorable à l'union civile de personnes de même sexe," *La Nouvelle République*, October 22, 2020.

375. *Sexo en piedra*, exhibition at the Fundacion Atapuerca, September 23 to December 8, 2017.

376. Florence Evin, "Pour qui donc, à Pompéi, s'élevaient ces phallus?," *Le Monde*, January 9, 2012.

377. Suetone, *Vie des douze Césars* (Paris: Le Livre de poche, 1961).

378. Bagemihl, *Biological Exuberance*.

379. Thierry Lodé, *La biodiversité amoureuse: Sexe et évolution* (Paris: Odile Jacob, 2011).

380. Geoff R. MacFarlane et al., "Same-sex sexual behavior in birds: Expression is related to social mating system and state of development at hatching," *Behavioral Ecology* 18, no. 1 (January 2007): 21–33.

381. Rebecca Kessler, "Why it's OK for birds to be gay," *Live Science*, August 23, 2010.

382. Bagemihl, *Biological Exuberance*

383. Goldfoot et al., "Behavioral and Physiological Evidence of Sexual Climax in the Female Stump-Tailed Macaque (*Macaca arctoides*), 1477–1479.

384. "Un couple de manchots gays adopte un petit avec succès," *Le Monde*, June 3, 2009.

385. MacFarlane et al., "Same-sex sexual behavior in birds, 21–33.

386. Lindsay C. Young, Brenda J. Zaun, and Eric A. Vanderwerf, "Successful same-sex pairing in Laysan albatross," *Biology Letters* 4, no. 4 (August 23, 2008): 323–325.

387. Charles Baudelaire, "L'Albatros," *Les fleurs du mal* (Paris: Flammarion, 2011).

388. B. Smuts and Robert W. Smuts, "Male aggression and sexual coercion of females in nonhuman primates and other mammals: Evidence and theoretical implications," *Advances in the Study of Behavior* 22 (1993): 1–63.

389. Randy Thornhill and Craig T. Palmer, *A Natural History of Rape: Biological Bases of Sexual Coercion* (Cambridge, MA: MIT Press, 2001).

390. Frans de Waal, "Et si l'homme descendait du viol?," *BibliObs*, January 9, 2012.

391. Joseph Henrich, Robert Boyd, and Peter J. Richerson, "The puzzle of monogamous marriage," *Philosophical Transactions of the Royal Society B: Biological Sciences* 367, no. 1589 (March 5, 2012): 657–669.

392. William A. Haddad et al., "Multiple occurrences of king penguin (*Aptenodytes patagonicus*) sexual harassment by Antarctic fur seals (*Arctocephalus gazella*)," *Polar Biology* 38 (November 2015): 741–746.

393. Mikkel Skovrind et al., "Hybridization between two high Arctic cetaceans confirmed by genomic analysis," *Scientific Reports* 9, no. 1 (June 20, 2019): 1–10.

394. Dr. Manhattan, "Nous ne vivons pas une sixième extincton de massé," *Rage*, January 3, 2020, http://rage-culture.com/nous-ne-vivons-pas-une-sixieme-extinction-de-masse/.

395. Philippe Mercure, "Un embryon hybride singe-humain créé en Chine," *La Presse*, August 10, 2019.

396. Nicolas Bastuck, "Zoophilie: 'Il existe un voyeurisme massif sur Internet,'" *Le Point*, January 23, 2020.

397. Marjolaine Baron, "La zoophilie dans la société: Quel rôle le vétérinaire peut-il tenir dans sa répression?" (PhD diss., Toulouse Veterinary School, 2017).

Epilogue. Reconciling with Our Animality

398. Florence Burgat, *L'humanité carnivore* (Paris: Le Seuil, 2017).